The Geopolitics of Energy into the 21st Century

The Geopolitics of Energy into the 21st Century

Volume 3: The Geopolitical Outlook, 2000–2020

A Report of the CSIS Strategic Energy Initiative

Project Cochairs
Sam Nunn
James R. Schlesinger

Project Director
Robert E. Ebel

Project Executive Director
Guy Caruso

Congressional Cochairs
Senator Joseph I. Lieberman
Senator Frank H. Murkowski
Representative Benjamin Gilman
Representative Ellen O. Tauscher

Rapporteur
Elaine L. Morton

November 2000

About CSIS

The Center for Strategic and International Studies (CSIS), established in 1962, is a private, tax-exempt institution focusing on international public policy issues. Its research is nonpartisan and nonproprietary.

CSIS is dedicated to policy impact. It seeks to inform and shape selected policy decisions in government and the private sector to meet the increasingly complex and difficult global challenges that leaders will confront in this new century. It achieves this mission in four ways: by generating strategic analysis that is anticipatory and interdisciplinary; by convening policymakers and other influential parties to assess key issues; by building structures for policy action; and by developing leaders.

CSIS does not take specific public policy positions. Accordingly, all views, positions, and conclusions expressed in this publication should be understood to be solely those of the authors.

President and Chief Executive Officer: John J. Hamre
Senior Vice President and Director of Studies: Erik R. Peterson
Director of Publications: James R. Dunton

Library of Congress Cataloging-in-Publication Data

The geopolitics of energy into the 21st century / a report of the CSIS Strategic Energy Initiative ; project cochairs, Sam Nunn, James R. Schlesinger ; project director, Robert E. Ebel.
p. cm. — (CSIS panel reports)
Includes bibliographical references.
Contents: Vol. 1. An overview and policy considerations.—Vol. 2. The supply-demand outlook, 2000–2020.—Vol. 3. The geopolitical outlook, 2000–2020.
ISBN 0-89206-367-X (set)—ISBN 0-89206-368-8 (v. 1)—ISBN 0-89206-369-6 (v. 2)—ISBN 0-89206-370-X (v. 3)
1. Energy policy. 2. Energy consumption—Forecasting. 3. Power resources. 4. Geopolitics. 5. International economic relations. I. Nunn, Sam. II. Schlesinger, James R. III. Ebel, Robert E. IV. Center for Strategic and International Studies (Washington, D.C.). V. Series.
HD9502.A2 G465 2000
33.79'01'12—dc21 00-029493

The CSIS Press
Center for Strategic and International Studies
1800 K Street, N.W., Washington, D.C. 20006
Telephone: (202) 887-0200
Fax: (202) 775-3199
E-mail: books@csis.org
Web site: http://www.csis.org/

Contents

List of Tables and Figures

Acknowledgments

The CSIS Strategic Energy Initiative (SEI) was launched in July 1998 in response to the concerns of former senator Sam Nunn and former secretary of energy James R. Schlesinger, who believed that a lengthy period of secure energy supplies had led U.S. policymakers to neglect the changing relationship between geopolitics and energy at the turn of the century. This changing relationship will require a rethinking of U.S. foreign policies, environmental policies, and the broader national security strategy. The three-volume study, *The Geopolitics of Energy into the 21st Century,** responds to these concerns.

Experts from the private and public sectors and from academe were brought together to discuss and analyze future energy supply and demand and to balance this analysis with a review of the key oil- and gas-exporting countries of the world. Guy Caruso took on the difficult task of day-to-day management of SEI, ably supported by Megan Smith in her position as project coordinator. Elaine O. Morton as rapporteur had perhaps the most difficult assignment, that of weaving together what is a complex story. She succeeded in every regard and is recognized as the principal author of all three volumes.

The project benefited immensely from the continued support and involvement of Anthony A. Smith, then CSIS executive vice president and chief operating officer, and Robin Niblett, then vice president for strategic planning. Their encouragement was essential to SEI's successful conclusion. Equal thanks are due the many CSIS program directors and other CSIS staff for their contributions to volume 3, *The Geopolitical Outlook:* Gerrit Gong, Teresita Schaffer, Simon Serfaty, Lowell Fleischer, Howard Wiarda, Anthony Cordesman, Judith Kipper, Keith Bush, Connie Freeman, and Steve Morrison.

A number of individuals offered their insight and judgments throughout the project. We particularly recognize Bill Hale, Arnold Baker, Wilfrid Kohl, Herman Franssen, and Jim Ragland. To each of the task force members, without whose enthusiastic participation SEI would not have been possible, and to Lisa Hyland, who took over upon the departure of Megan Smith, we extend our personal thanks.

Special recognition must be given to congressional cochairs Senators Frank H. Murkowski (R-Alaska) and Joseph I. Lieberman (D-Conn.) and Representatives Benjamin Gilman (R-N.Y.) and Ellen O. Tauscher (D-Calif.). Their continued support, and that of their staff, was essential to a successful SEI.

---

* The three-volume study includes *An Overview and Policy Considerations* (volume 1), *The Supply-Demand Outlook, 2000–2020* (volume 2), and *The Geopolitical Outlook, 2000–2020* (volume 3). Although the SEI Advisory Committee has reviewed volume 1, responsibility for the contents rests with CSIS, as it does for the two companion volumes.

Finally, the remarkable leadership of cochairs Sam Nunn and James R. Schlesinger throughout the project must be noted. Their personal involvement and guidance have been of tremendous importance.

CSIS is deeply appreciative of the efforts of all participants to produce a report that is timely, objective, and bipartisan and that carries policy considerations serving the best interests of the United States and of energy producers and consumers everywhere.

The geopolitics of energy is rarely static. Events of the day carry implications for energy supply, consumption, and prices—sometimes immediate, sometimes delayed, sometimes hidden. By attempting to define, in advance, the form that these events might take—and the resulting impact on energy—this report may remove some surprise from the future and ease the way for decisionmakers in both the public and private sectors.

Robert E. Ebel
Director, Energy and National Security
CSIS

Participants

Project Cochairs

Sam Nunn
James R. Schlesinger

Congressional Cochairs

Senator Joseph I. Lieberman (D-Conn.)
Senator Frank H. Murkowski (R-Alaska)
Representative Benjamin Gilman (R-N.Y.)
Representative Ellen O. Tauscher (D-Calif.)

Advisory Committee

Rodney F. Chase
Deputy Group Chief Executive
BP

Charles Fabrikant
Chairman, President & CEO
SEACOR SMIT, Inc.

Luis Giusti
Former Chairman and CEO
Petroleos de Venezuela, S.A.

Richard K. Gordon
Vice Chairman, Investment Banking
Merrill Lynch

Robert L. Healy
Vice President, Federal Government Relations
ARCO

Dale P. Jones
Former Vice Chairman
Halliburton Company

Diana MacArthur
Chair and CEO
Dynamac Corporation

Jessica T. Mathews
President
Carnegie Endowment for International Peace

John V. Mitchell
Chairman, Energy and Environment Program
The Royal Institute of International Affairs

David M. Nemtzow
President
The Alliance to Save Energy

Johan Nic Vold
Executive Vice President
Statoil

C. Paul Robinson
President
Sandia National Laboratories

William Ruckelshaus
Former Chairman
Browning-Ferris Industries

John W. Snow
Chairman, President & CEO
CSX Corporation

Robert E. Wilhelm
Director and Senior Vice President (retired)
Exxon Mobil Corporation

Daniel B. Yergin
President
Cambridge Energy Research Associates

Program Director

Robert E. Ebel

Project Executive Director

Guy Caruso

Rapporteur

Elaine L. Morton

Project Coordinator

Megan Smith

Task Force Members

A total of eight task forces were assembled, from experts in both the private and public sectors, to carry out the work of the Strategic Energy Initiative project. Task forces were established on Asia, the Middle East, the former Soviet Union, the Western Hemisphere, Africa, Europe, the Environment, and Technology. Although many of the experts served on several task forces, their names are presented here in a combined list so as to avoid redundancy. In addition, those participants who offered their insights on a broader basis and were not members of any particular task force are also recognized.

Art Andersen
U.S. Department of Energy

Kai Anderson
Office of Senator Joseph Lieberman

Arnold Baker
Sandia National Laboratories

David D. Bosch
Aramco Services Company

John Brodman
U.S. Department of Energy

Jim Bruce
Wiley, Rein, & Fielding

Todd Burger
Office of Representative Benjamin Gilman

Ellison Burton
Dynamac Corporation

G. Daniel Butler
U. S. Department of Energy

Alys Campaigne
Office of Senator Joseph Lieberman

Michael Canes
American Petroleum Institute

Galen Cobb
Halliburton Company

Len Coburn
U.S. Department of Energy

Melissa Coffey
Office of Representative Benjamin Gilman

Barry Cohen
Energy Information Agency

Bud Coote
Central Intelligence Agency

Terry Coyne
U.S. Government

Peter A. Davies
BP

Ged Davis
Shell International Limited

Fred Downey
Office of Senator Joseph Lieberman

Luther W. Dudich
Alliance to Save Energy

Charles Ebinger
International Resource Group

Lana Ekimoff
U.S. Department of Energy

Marvin Fertel
Nuclear Energy Institute

George Folsom
Koch Industries

Herman Franssen
International Energy Associates

Chas Freeman
Middle East Policy Council

Steve Gallogly
U.S. Department of State

Larry Goldstein
Petroleum Industry Research Foundation

Robert L. Gould
CSX Corporation

William Hale
Exxon Mobil Corporation

Dan Hickey
Texaco

Lynn Hogan
Arco

Karen Hunsicker
Senate Committee on Energy and Natural Resources

Maureen Koetz
Nuclear Energy Institute

Wil Kohl
SAIS
Johns Hopkins University

Michael V. Kostiw
Texaco

Peter Lyons
Office of Senator Pete V. Domenici

Neil McKeown
U.S. Government

Charles McPherson
The World Bank

Eric Melby
Scowcroft Group

Jack Moore
Aramco Services Company

Angela K. Morin
Alliance to Save Energy

Ed Morse
Hess Trading Company

Steven Mullen
U.S. Department of State

Peter Muller
Office of Representative Ellen Tauscher

Shirley Neff
Senate Committee on Energy and Natural Resources

Deanna Okun
Office of Senator Frank Murkowski

Willy H. Olsen
Statoil

Michelle R. Perez
Alliance to Save Energy

James Placke
Cambridge Energy Research Associates

Jerry Pruzan
Arco

James Ragland
Aramco Services Company

Joyce Rechtschaffen
Office of Senator Joseph Lieberman

Richard Sawaya
Arco

Robert Shinn
CSX Corporation

Adam L. Shrier
International Consulting

Adam E. Sieminski
BT Alex Brown

Ron Simard
Nuclear Energy Institute

Elizabeth A. Sollie
Texaco

Vito Stagliano
Commonwealth Edison Company

Cyrus Tahmassebi
Energy Trends, Inc

Michael Townshend
BP

Geir Westgaard
Statoil

Guenther O. Wilhelm
Exxon Corporation (retired)

Abbreviations

The following abbreviations are used in the three-volume study.

Bbbl	billion barrels
bbl	barrels
Bcf/d	billion cubic feet per day
Bcm	billion cubic meters
BkWh	billion kilowatt hours
Bst	billion short tons
Bt	billion metric tons
Btu	British thermal unit
kW	kilowatt
kWh	kilowatt hour
MMbbl/d	million barrels per day
MMst	million short tons
MMt	million metric tons
mpg	miles per gallon
MW	megawatt
quad	quadrillion Btu
st	short ton
t	metric ton
Tbbl	trillion barrels
Tcf	trillion cubic feet
Tcm	trillion cubic meters

Numbers used in this report are defined according to the system used in the United States.

million	10^6 (6 zeros)
billion	10^9 (9 zeros)
trillion	10^{12} (12 zeros)
quadrillion	10^{15} (15 zeros)

CHAPTER 1

Introduction

The Challenge

As this is written, in 2000, more than 20 years have passed since world energy markets experienced the dramatic oil supply disruption and sharply increased oil prices caused by the Iranian revolution.[1] Almost 10 years have passed since the oil supply disruption and oil price increase caused by Iraq's invasion of Kuwait.

In 2000, however, the condition of the world energy sector has been characterized by rapidly rising oil prices as oil-exporting countries curtail production; this contrasts with the ample supply and low-to-moderate prices for most of the 1990s. Energy consumers in earlier months had experienced unexpected benefits, benefits followed, however, by the pressures that high prices bring in their wake.

Meanwhile, reliance on markets and interdependent relationships have accelerated in the economic realm, and political pluralism and democratization have accelerated in the political realm.

The passage of time and the political and energy supply situation at the turn of the twenty-first century nonetheless have produced an atmosphere of relative calm as far as energy security is concerned—a striking contrast with the widespread perception of an energy crisis during the 1970s and early 1980s.

This sense of calm masks two dangers:

- It has obscured certain underlying shifts in the geopolitics of future global energy security, supply, and demand; and
- Policymakers and the public at large can become complacent about the importance of considering global energy issues when foreign policy decisions are made.

With these factors in mind, CSIS in July 1998 launched its Strategic Energy Initiative (SEI) designed to examine the relation between energy and geopolitics—as it exists now at the beginning of the twenty-first century, and as it is likely to unfold during the next two decades.

This analysis will provide the basis for identifying risks and opportunities that are deserving of international policy attention and consideration.

---

1. The supply interruption caused by the Iranian revolution resulted in high oil prices that were later capitalized on by an Organization of Petroleum Exporting Countries (OPEC) strategy to keep the prices at high levels by reducing OPEC members' output.

Energy and Politics: A Key Relationship

The relation between energy and politics is a strong one in developed and developing countries alike.

- In developed countries, standards of living and life style expectations are heavily dependent on high levels of energy use. At a national level, economic power—based in large part on highly sophisticated use of energy resources—translates into geopolitical power on the world stage.
- In developing countries, particularly those moving rapidly to higher standards of living and industrialization, energy availability is key to whether national goals for development will be met, whether the expectations of local populations can be satisfied, and whether national aspirations to greater political power in the international arena can be pursued effectively.

Energy and Geopolitics: A Reciprocal Relationship

Geopolitics and Energy...

The potential impact of geopolitical developments on energy availability is not difficult to envision. Producers and consumers can reflect on the experience of the OAPEC oil embargo[2] imposed in response to U.S. support of Israel during the October 1973 war as well as the sharp increases in oil prices that occurred during the Iranian revolution in 1979, during the initial phase of the Iran-Iraq War in 1980, and immediately following Iraq's invasion of Kuwait in 1990. It is clear that any major conflagration in the Middle East would almost certainly lead to a dangerous oil supply interruption—either by accident or by the intention of one of the parties to the conflict.

In the altered post–Cold War period, moreover, where the disappearance of the East-West standoff has weakened outside restraints on the escalation of regional conflicts, political instability affecting major oil exporters may prove far more difficult to control.

Concurrent with the end of the Cold War constraints, there has been an explosion in cheap, accessible communications, including the Internet. Advocates of issues such as human rights and environmental protection can organize cheaply and communicate news and ideas extensively across frontiers to influence public opinion and, therefore, policies in the United States and other major countries.

---

2. The Organization of Arab Petroleum Exporting Countries (OAPEC) initiated and implemented the 1973 oil embargo that—in contrast to popular misconception—was not an OPEC undertaking. OPEC is scrupulous about not taking any positions on political issues and has a membership that includes countries (like Venezuela and Nigeria, for example) that are not located in the Middle East and do not share the political concerns of its Arab members. OAPEC, on the other hand, includes Arab countries that are also members of OPEC along with non-OPEC members like Egypt, Oman, and Syria.

...Energy and Geopolitics

The potential impact of energy issues on geopolitical outcomes is not as widely appreciated as is the reverse relation. Nevertheless, the effect of energy issues may achieve greater salience in the years to come. Examples of this interrelation include

- The mid-1990 oil price dispute between Kuwait and the United Arab Emirates (UAE) and Iraq that became an important precipitating factor in Iraq's ultimate decision to invade Kuwait.
- The 1991 Gulf War itself. Although the primary articulated objective of the United States and allied forces came to be the liberation of Kuwait and restoration of its sovereignty, the issue of energy security was also a powerful motivating force. If he had not been turned back, Saddam Hussein would have been able to control two major oil-producing countries (Iraq and Kuwait) while he exerted pressure on a third (Saudi Arabia).
- Price fluctuations, deriving from supply-demand imbalances though mitigated somewhat by substantial tax structures, can and do have an impact on the stability of key energy-importing countries in Europe. For years to come, the states of Europe will need an especially benign economic environment. Europe's ability to reform, widen, and deepen its institutions may depend on growth rates closer to 3 percent and above than to 1 percent or below. Under such conditions, Europe's dependence on oil imports at stable prices may be even more significant in the coming two decades than it was in the 1970s, when two oil crises derailed the European Community. At the same time, in 2004 after the euro becomes the sole tender for most or all European Union (EU) states, European vulnerability to price fluctuations could be reduced if prices are denominated in euros as well as dollars.
- More recently, U.S. policymakers hope that rapid and effective development of energy resources in the Caspian Sea will enable the Caspian region to achieve economic and political independence from Russia. Unfortunately, for a variety of reasons, development has not been as rapid or as effective as initially hoped.
- Less obvious, but also important, is the paradox that the very existence of abundant energy wealth in some developing countries can be a factor in promoting political instability or sustaining political conflict.
- Unrest in the Niger Delta area of Nigeria is one example in which perceived corruption in the use of oil income and inequality in regional distribution of oil income have led to political protest and internal violence.
- In neighboring Angola, a devastating 30-year civil war has been funded largely through oil and mineral wealth.
- Access to oil revenues enables authoritarian regimes to avoid the public accountability and the political flexibility that democratic processes can provide.
- It is not inconceivable that extreme conditions in the world energy market could cause future geopolitical repercussions.

- A glut of supplies and fierce competition between exporting countries for market share could lead low-cost producers to try to drive high-cost producers out of the market, or could even lead to military conflict as a means of trying to thwart rivals and ensure market supremacy. This is possible especially in circumstances in which oil producers are dangerously dependent on oil income to ward off budget default or internal instability.
- Extreme scarcity of energy supply and competition between major consumers could similarly be expressed in political and military ways.
- A severe, albeit temporary disruption of oil supplies with a 3–6 month price hike could provoke either a burst of world inflation or, if restrictive policies were adopted, possibly a severe recession.
- A prolonged failure to invest in new, long-lead-time energy supplies could affect world economic growth through inflated energy prices and a boom-and-bust investment cycle in energy-related capital.
- Finally, one of the issues that this report will examine in more detail is the way that increasing energy use in China and Asia during the coming two decades might affect more broadly the geopolitical relations in the region.

The Time Dimension: 2000 to 2020

The challenge of projecting the complicated interrelationship of energy and geopolitics over a time span of two decades is daunting. Although estimates can be made by analyzing trends and known relationships, many uncertainties and important events—discontinuities—cannot be predicted at all.

A Backward Look at Change

A useful way to begin thinking ahead about the variety of things that could have an impact on the geopolitics-energy equation during the years 2000–2020 is to look backward and recall some of the key energy-related events that have occurred during the most recent 20-year period. These include

- The 1980–1988 Iran-Iraq War. Oil tankers were attacked, Kuwaiti tankers received U.S. protection by being reflagged, and Kharg Island (a major oil export terminal located in the upper Gulf) was bombed.

  According to the International Energy Agency (IEA), at the outset of the war 6 percent of the oil supply to the world market was cut off, and the disruption lasted for 4 months. During the remaining seven years of the war, neither Iran nor Iraq contributed fully to world oil supply.

- The 1985–1986 oil price crash caused by Saudi Arabia's decision to lower its prices in order to gain a larger share of the market.[3] Prices fell from $28 per barrel in 1985 to lows of $6–$9 per barrel during the summer of 1986. In 1987, when OPEC returned to a policy of production restraint, prices stabilized at $18 per barrel—a 36 percent drop from the price level of 1985.[4]

- The collapse of the Soviet Union oil sector in 1988. Once the world's largest oil producer, Soviet production declined drastically, to barely one-half of 1988 levels 10 years later.
- The 1990–1991 dissolution of the Soviet Union, which permitted external access to its former Caspian energy reserves.
- The August 1990 Iraqi invasion of Kuwait. UN economic sanctions—including an oil embargo—were imposed in reprisal. The global oil market suffered an abrupt 4–5 MMbbl/d loss of supply immediately thereafter.

  Subsequently, Saudi Arabia, the UAE, Venezuela, and other countries made up the shortfall. The Saudis alone were able to contribute a production increase of 2 MMbbl/d.
- In 1993, China—a rapidly industrializing country with 1.2 billion people—became a net oil importer, with prospects for rapidly growing oil imports and a significant impact on the world oil market.
- During the 1990s, technological advances in oil exploration and production opened increasingly deep offshore areas to exploitation and contributed to higher levels of both oil recovery and oil supply.
- But, during the past two decades, new discoveries of oil have waned. Growth in world oil reserves has largely been brought about instead by higher recovery factors and reappraisal of reserve levels.
- Meanwhile, nuclear power increased its contribution to worldwide energy supplies—but the building of new nuclear plants has ceased in the United States (in Europe, some governments plan to retire existing plants).
- In December 1997, the Kyoto Protocol[5] was signed. Most Organization of Economic Cooperation and Development (OECD) countries ostensibly undertook commitments to reduce their carbon emissions below 1990 levels by 2012.
- Between mid-1997 and early 1999, world oil prices dropped to their lowest level in real terms since before the 1973 oil embargo because of unusual decreases in demand paralleled by increased supply.[6]
- But by mid-1999 oil prices had doubled, a reversal accomplished by the combined action of OPEC and several other oil exporters to curtail supplies as well as the movement of demand toward previous higher levels. Prices continued to

---

3. Note that Saudi Arabia was not alone among OPEC members increasing production to gain market share. "From the third quarter of 1985 to the third quarter of 1986, the Saudis expanded output by 131 percent, Kuwait by 69 percent, Iraq by 50 percent, and the United Arab Emirates (UAE) by 38 percent." Among the OPEC members that did not have the ability to surge their production levels, oil revenues declined simultaneously in 1985–1986: 60 percent for Algeria, 49 percent for Nigeria, 40 percent for Indonesia, and 50 percent for Iran. Data presented in Edward N. Krapels, "The World Oil Market of the 1980s," in *After the Oil Price Collapse: OPEC, the United States, and the World Oil Market*, ed. Wilfred L. Kohl (Baltimore: Johns Hopkins University Press, 1991), 52.

4. Fadhil J. Chalabi, "The World Price Collapse of 1986: Causes and Implications for the Future of OPEC," in *After the Oil Price Collapse*, 1.

rise during the remainder of the year and into 2000, exceeding $30 per barrel in March 2000.

Few of these events were foreseeable, which can only underscore the geopolitical uncertainties that lie ahead.

Looking Forward

With events and developments like these occurring during the past 20 years, it is not difficult to imagine that the next 20 years will offer similarly important changes.

Some of the future events will be variations on past themes. Saudi Arabia may follow a price defense rather than a market-share strategy, for example. The Caspian may emerge as a major oil and gas province. Fallout from the Gulf War will continue in the form of further weakening or removal of sanctions against Iraq. The degree to which China enjoys energy security will be important. The politics and economics of climate change will continue to unfold.

Some of the changes will be surprising and impossible to predict. A supply interruption—natural or man-made—could occur, threatening the security and economic interests of the United States and all other countries, whether in the industrialized or developing worlds.

Conversely, the coming years may witness little more than sporadic price fluctuations, with only temporary impact on economic activity.

In sum, all of the changes that will be experienced will be managed best if policymakers make an effort today to avoid the most extreme contingencies and prepare for those that cannot be deflected.

Policy Focus

Countries of the world are now increasingly engaged with each other in a global context. This is particularly true when it comes to energy. Not only is the energy market itself global in nature, but international trade, more generally, can be envi-

---

5. This refers to the Kyoto Protocol of the 1992 Framework Convention on Climate Change (FCCC; the Rio Treaty). The United States Senate must ratify the Kyoto Protocol for it to have a binding effect on the U.S. government. In addition, the terms of the protocol itself stipulate that it will not come into effect until it is ratified by 55 countries, including countries that together contribute 55 percent or more of the world's carbon emissions. Because the United States is judged to contribute 22–24 percent of the world's carbon emissions, the protocol theoretically could come into effect in international law without U.S. ratification if other countries with significant levels of carbon emissions sign it. The 22 percent figure comes from *Climate Change; State of Knowledge* (Washington, D.C.: Office of Science and Technology Policy, October 1997), 6. The 24 percent figure comes from *Climate Protection Policies: Can We Afford to Delay?* (Washington, D.C.: World Resources Institute, 1997), 18.

6. The sharp decline in demand was caused by continuation of an economic slowdown in Asia, coupled with an unusually warm El Niño winter. The sharp increase in supply was caused partly by OPEC's November 1997 decision to increase oil output quotas and partly by relaxation of the terms of the oil-for-food plan that the United Nations had created for Iraq. Under the relaxed restrictions, Iraq raised production to 2.2 MMbbl/d—an 85 percent increase from its 1997 level.

sioned as consisting of imports and exports of goods, each of which has a specific energy content.

Therefore, even if it were hypothetically possible for a particular country to achieve total energy self-sufficiency, that country would still be dependent on the ability of other countries to

- Acquire enough energy to continue to produce exported goods, and
- Engage in energy-sustained economic activity at levels that would provide sufficient income for continuing to acquire imported goods.

Otherwise, the ability of the world economic system to exploit the gains that accrue from comparative advantage in production would break down.

It is for this and other reasons that CSIS's SEI is focused not only on the objective of achieving energy security for the United States. The objective is achieving international energy security.

In addition, producers and consumers should work toward an international energy system that makes use of market efficiencies while it enables governments and private industry to address problems that the market alone cannot solve.

Such a world energy system would provide sufficient supplies of energy—at reasonable cost—in order to enable

- Developed countries to continue to enjoy and improve their current standards of living, and
- Developing countries to improve their standards of living while achieving sustainable economic development in the process.

Because of the interrelation between energy and international politics, it is also important to state the collateral objectives of mitigating

- The outbreak of international conflicts that could cause significant interruptions of energy supply,
- The possibility that energy issues will precipitate instances of international conflict in their own right,
- The possibility that energy issues will exacerbate the potential for domestic instability in energy-producing countries,
- Situations of domestic instability in energy-producing countries that interfere with production and export capabilities.

Finally, one of the most important objectives of this study is to stress the need to construct policies that will take as a given the world's continuing dependence on fossil-fuel energy sources while they simultaneously fulfill the imperative to address the collateral impact on the world's environment.

Although environmental issues are important, the forces of geopolitics have not been suspended and their importance is greater than their reflection in the current energy debate.[7]

Policy Goals

The policy considerations of this study will be grouped in categories that delineate the following basic policy goals:

1. To ensure the adequacy of future international energy supplies,
2. To ensure reliable and timely access to these supplies, and
3. To pursue energy policies that protect the environment.

Providing a Strategic Approach

The purpose of the SEI is to help policymakers plan ahead strategically by stressing the importance of preparing for contingencies. The range of challenges and opportunities that policymakers face will depend on developments in the field of energy and on the political and economic choices that other nations make. Focusing on the longer term, this report hopefully will help policymakers better understand the long-term implications of near-term policy decisions.

Although for ease of presentation it is often convenient to present options for U.S. policymakers, the SEI purpose is broader in scope. Situations described here and analytic conclusions reached can be of use more generally to members of the international community and to both the public and private sectors—each with its own separate perspective.

---

7. In the United States, for example, the Executive Office of the President has delegated energy issues to the staff of the National Economic Council, where they are dealt with in domestic rather than international terms. In previous administrations, energy issues were considered an important aspect of international relationships, and the staff of the National Security Council held the energy portfolio.

CHAPTER 2

The Geopolitical Outlook

Overview

Several aspects of the geopolitical situation in 2000 are important in conditioning the way in which energy issues and relations will unfold during the time covered by this study. Among these, the change in international political relations caused by the end of the Cold War is most prominent. But several broad global trends—both economic and political—also have an impact on the way in which energy is created and used.

- From an economic standpoint, the most important of these trends is a growing reliance on free trade and private sector solutions for resolving a host of problems that were formerly considered to be within the province of governments. This has happened within the liberal democracies, through measures to privatize and deregulate. It has also happened within a larger context—with the collapse of communism in the countries of the former Soviet Union, the continuation of China's policy of economic reform, and the abandonment of the socialist model in almost all of the countries of the world.

  Moreover, many of the major energy-producing countries by necessity are pursuing policies that reflect a significant decline of resource nationalism. Because new investment opportunities have opened up, the oil industry is no longer as opportunity constrained as it was during the 1970s and 1980s.

- From a political standpoint, a global environmental movement has gained significant momentum, as manifested by the ratification of the FCCC and the Kyoto Protocol as well as by a host of environmental objectives currently being pursued in national and local forums.

  Although some observers might dissent, the nexus of environmental and energy issues is so strong that the EIA's *International Energy Outlook 1999* asserts that

  > the principal international energy issues have shifted from supply interruptions and their implications for energy security and price stability to the impact of energy production and consumption on regional and global environments.[8]

  In addition, there has been growth in the political activism of private groups,[9] which has made corporate social responsibility more of an issue in the business

8. *International Energy Outlook 1999* (Washington, D.C.: Energy Information Administration, U.S. Department of Energy, March 1999).

9. Private groups are usually referred to as nongovernmental organizations, or NGOs.

world than it used to be. Today, more than ever before, perception equals reality. Corporations consequently guard their public images carefully. At the same time, new information technology has created a knowledge-based society in which activists are empowered to alter public perceptions in order to effect political change.

- From the standpoint of political economy, the geopolitical landscape is marked by significant differences among countries in terms of obvious trends of population growth, urbanization, and economic development. Each of these trends contributes to increased levels of global energy demand. The balance among them in any individual country substantially influences the level of political stability that a country will enjoy.

  In particular, the degree of political stability that characterizes the major developing-country energy exporters is an important variable in determining the likelihood of uninterrupted supply to global energy markets. The countries of the Persian Gulf, for example, have population growth rates that exceed 3 percent per year. The full impact of the population increase will be experienced when the newly born reach adulthood. This population crunch will occur simultaneously with two other events: succession crises for many of the region's ruling families and a growing realization that high oil revenues cannot be counted on indefinitely.

- From a military and security standpoint, the internal instability of many major energy exporters poses risks of supply interruptions and causes doubts about the ability of such countries to maintain and expand their energy exports. Advances in weaponry and military technology have increased the vulnerability of shipping routes around the world. This is especially true of the handful of sea-lane choke points[10] that mark the major routes of energy transport. At the same time, the gulf presents special problems because of the risks posed by potentially aggressive regimes in Iran and Iraq and the prospect for development and use of weapons of mass destruction (WMD).

Changed International Power Relationships

International power relationships at the turn of the twenty-first century differ broadly from those of the not-too-distant past in several major ways. This section examines three broad trends:

- The changed geopolitical landscape that has occurred with the end of the Cold War.
- The extent to which nation-states are being weakened by ethnic separatism and by growing political power wielded by international groups of private activists, and

10. See appendix A for a detailed description of choke points.

- The extent to which the power of weaker nation-states and separatist movements has been increased by their ability to use terrorism to influence events and by weapons proliferation that includes long-range ballistic missiles and chemical, biological, and nuclear capabilities—the so-called WMD.

Although the risk of the kind of global conflict seen during the twentieth century is dramatically reduced, a greater risk of international conflict will exist for several reasons.

The Geopolitical Landscape

Geopolitical power relationships have changed in many ways. The United States has emerged from the Cold War as, some say, "the world's only superpower," the only state capable of projecting massive amounts of conventional military power practically anywhere in the world. The preeminence of U.S. military power has led to new challenges. These include a growing resentment on the part of key powers like Russia and China, and many developing states, of what they consider is U.S. hegemonism. In this environment, many regionally preeminent nations and others see the possession of WMD as the only way to block U.S.-led intervention in regional struggles.

In addition, the reasserted role of the U.S. Congress in making U.S. foreign policy exposes Congress to special-interest groups and single-issue advocates whose effectiveness is enhanced by the use of the media, the Internet, and cheap and accessible communications. The Iran-Libya Sanctions Act, with its implications, was seen by many outside the United States as an exercise in extraterritorial hegemony without accountability.

Power is shifting in other ways. New centers of economic power are developing; these may ultimately prove to be at least as important as the centers of military power that shaped power relationships during the Cold War. The new centers include the emergence of the EU and the development of Asia with its major economic centers in China and Japan as forces in the world economy.

At the same time, the end of the Cold War has led to a steady growth of regionalism, in the sense that the disputes and conflicts taking place in the world increasingly focus on local, national, and ethnic issues. Apart from the Middle East and, perhaps in the future, Central Asia and the Caspian, the new areas of potential instability do not directly affect energy supplies to the United States and Europe. The political and economic effects of instability in the energy-poor areas would far outweigh any energy implications.

However, commitment of U.S. resources to stabilize developing political turmoil in these areas will limit U.S. capacity to intervene also in the area of direct energy concern—the Middle East. Put simply, if U.S. military power were committed even to an extended protective exercise in Northeast Asia, the capacity to respond to a crisis like that of 1990 in the Middle East would be severely limited.

At the same time, many U.S. allies are concerned that in the absence of shared Cold War priorities, U.S. policy might become less committed to regional security, and U.S. military engagement might be less likely in conflicts that mainly affect Europe or Northeast Asia. For the same reasons, there is more scope for difference in objectives and tactics between the United States and its European and Asian allies regarding areas like the Middle East and Central Asia.

- For Europeans, an evolving answer to these geopolitical changes is the development of a common European foreign and security policy, ideally compatible with NATO membership but with an institutional base in the EU and its own military forces.
- For U.S. allies in Northeast Asia, the position is more difficult. Unlike the countries of Western Europe, U.S. allies in Northeast Asia face potential direct threats to their own integrity from North Korea and China. Regional security essentially continues to depend on bilateral relationships with the United States and, therefore, on the nature and degree of U.S. commitment. Changes in the scope of Japanese "self-defense" preparedness do not alter the geopolitical dependence of these countries on the United States.

  In matters of energy, however, the broadening scope of the Asia Pacific Economic Cooperation (APEC) and the establishment of an energy working group under the APEC umbrella offer the basis for developing common approaches to energy policy and security—short of a military dimension. These may also evolve to provide a basis for constructive engagement—to use the European term—with Middle Eastern oil-exporting countries. As with the European dialogue, this may lead to energy-related initiatives independent of U.S. policy.
- In South Asia, the explicit emergence of India and Pakistan as nuclear-weapons states raises profound questions: Will they be able to resolve their disputes and secure regional peace? Will other countries also go down the nuclear path?

Political Power of Nation-States

An increased tendency of local populations to identify with separate ethnic groups has caused a limitation of internal power within nation-states. At the same time, and for different reasons, nation-states and treaty-based international organizations have begun to share political decision-making power with a variety of private NGOs.

Ethnic Separatism

Ethnic separatism and conflict are major sources of political instability and cross-border warfare. It is not entirely clear why identification with ethnic groups has intensified in recent years—even in developed, industrialized countries like the United States.

One explanation is that the stresses imposed by rapid change from traditional to modern ways of life in many parts of the world have reinforced the need for people to bond together for a feeling of security in groups that are closer to home than the nation-state.

Another explanation is that national governments in advanced technological societies have become somewhat remote from the everyday concerns of their constituents and, again, it is more satisfactory to search for political solutions in places closer to home.

Ethnic conflicts and failed political and economic development can interact to create civil wars and regional conflicts that affect energy-producing and transit countries. Some of the major energy exporters in the developing world are at least potentially at risk.

Private Activism

Groups composed of citizen-activists increasingly have been formed to address policy issues that are either global in nature or involve the internal affairs of other states. The NGOs' geopolitical significance derives from the extent to which they have been able take effective action in policy areas that formerly were the sole province of governments and treaty-based international organizations.

Meanwhile, the global spread of cheap communications and the influence of the visual media increasingly enable disaffected groups within states to reach for support outside their borders. Within the great democracies, especially the United States, excellent communications enable advocates to introduce single issues into the political and economic processes if national values—distinct from national interests—are involved.

The Internet, facilitating coordination and enhancing the ability to get a message out to an unusually wide audience, enables issue advocates to join forces across national boundaries. Funding from private donors, foundations, and even governments is often plentiful, and relatively little of this money has to be diverted to support an institutional infrastructure.[11] Technical expertise is either contributed or ready for purchase, and single-issue focus permits concentration of effort and effect.

Human Rights Issues. In the human rights field, private organizations from developed countries, well versed in lobbying tactics and media strategy, can link up with local groups of activists in developing countries, thereby enhancing the power of the local groups with respect to their home governments. International public opinion can be brought to bear by publicizing specific situations or events. Present and potential examples include sanctions against Burma (now Myanmar), a potential gas exporter; and the campaigns of Human Rights Watch to focus attention on human rights conditions in the oil-producing Niger Delta region of Nigeria and in oil-producing countries of the Caspian. It is possible that similar public attention could be aroused and focused on major Middle Eastern oil-exporting countries because of an absence of civil rights for women and lack of democratic institutions.

Close to 40 percent of the world's oil supply is produced in countries that, as of 1999, had not signed or ratified the main United Nations (UN) human rights covenants or were subject to serious criticism by the U.S. State Department and human rights organizations for inadequate protection of human rights.[12]

---

11. Jessica Mathews, "Power Shift," *Foreign Affairs* 76, no. 1 (January/February 1997): 55.

12. John Mitchell, ed., *Companies in a World of Conflict* (London: Royal Institute of International Affairs and Earthscan, 1998), 249–250.

In a new and significant development, some private NGOs have begun to use the U.S. courts to try to extend corporate liability to include alleged complicity in human rights abuses abroad.

- In 1996, a lawsuit was filed in a Los Angeles federal court in an attempt to hold Unocal liable for human rights abuses allegedly perpetrated by the Burmese army while it provided security protection for Unocal's offshore Yadana natural gas field project.
- In June 1999, a similar lawsuit was filed in federal court in San Francisco, alleging that Chevron is liable for the 1998 deaths of two Nigerian protesters because they were killed by Nigerian military police forces that Chevron had called in to assist in ending a disturbance at its Parabe offshore platform in the Niger River Delta.[13]

ENVIRONMENTAL ISSUES. NGOs played an especially prominent role in bringing about the negotiations that led to conclusion of the global climate treaty reached in Rio de Janeiro in 1992, which was followed later by the implementation measures outlined in the Kyoto Protocol of 1997. In fact, NGOs were intimately involved in the negotiation process itself.[14]

Environmental issues are intertwined with many aspects of energy exploration, production, and transportation—whether it is a question of protecting wildlife in Alaska, the coastline of Santa Barbara, the ecology of the area surrounding a proposed oil pipeline in Chad, or the offshore areas of Sakhalin Island. The role of private activists who have concerns about the environment is an important component of the business environment that energy companies face. Not only can treaties be signed and sanctions enacted in response to NGO activity, but boycotts can be organized, as they were against Shell over the *Brent Spar* issue.[15]

Increased Power of Weak States and Separatists

Trends toward increased proliferation of weapons and more frequent incidents of terrorism affect power relationships among nations.

Terrorism as an Instrument of Power

In a world in which it is common to measure international power in terms of military and economic strength, individual acts of extremism and terrorism have the power to influence events to an extent far beyond the actual physical destruction they cause. Terrorism is the weapon of choice of nations or groups that find themselves in a position of inferior strength vis-à-vis their adversaries.

Terrorism has had two effects on energy availability.

---

13. Gautam Malkani, "U.S. Groups Come under Spotlight over Human Rights Abroad," *Financial Times*, July 7, 1999. In the Chevron case, the police used helicopters and pilots contracted by the company, and the company's local security chief was on board.

14. Jessica Mathews, "Power Shift," 55.

15. The *Brent Spar* was an oil platform that was slated for disposal in a fjord near Norway. Environmentalists opposed that. Boycotts similar to that against Shell for the *Brent Spar* were placed against Shell and CalTex for conducting business in South Africa during the period of apartheid.

- First, several energy-exporting countries—as well as potential pipeline transit countries—are currently subject to U.S. sanctions[16] because of U.S. concern about terrorism.

  Of the seven countries that the U.S. secretary of state has designated as engaging in a pattern of support for terrorism,[17] five (Libya, Iran, Iraq, Syria, and Sudan) are energy producers; three (Libya, Iran, and Iraq) are major contributors to the world oil market; and two (Iran and Iraq) together possess close to 20 percent of the world's proved oil reserves.[18]

- Second, occurrences of terrorist-motivated sabotage can put oil and gas pipelines in temporary jeopardy, as often happens in Colombia and Algeria and has occurred at least one time that affected pipeline supplies from Algeria to Western Europe.[19] Nonetheless, these disruptions have had only local impact and damage has been quickly repaired; they have not yet influenced worldwide supply to any measurable degree. Today, infrastructures, including pipelines, face a new threat—cyberterrorism—for which an acceptable defense has yet to be found.

Weapons Proliferation as Diffusion of Power

It is clear that the perceived power of several regional states is being enhanced by the acquisition—or attempted acquisition—of advanced long-range missile capability and WMD. The weapons proliferation issue has several important implications for international energy relationships.

- The use of WMD in any military confrontation between countries in the Middle East or Northeast Asia could interfere with important patterns of energy flow.[20]
- Weapons proliferation has been one of the concerns that has led to sanctions policies that inhibit the expansion of crude oil productive capacity in countries with major energy resources.

---

16. The sanctions do not keep the oil exports of these countries out of the world market. Instead, because oil is fungible, sanctions merely rearrange the geographic pattern of its flow.

17. U.S. Code: Title 22, Section 2371. Countries listed as repeatedly providing support for acts of international terrorism are prohibited from receiving U.S. foreign assistance under the terms of Section 620(a) of the Foreign Assistance Act of 1961. Therefore, they cannot receive development assistance, economic support funds, or agricultural assistance under the PL480 program; and they cannot participate in Peace Corps programs or benefit from Export-Import Bank credits. Moreover, under Section 620(g) of the act, third countries that give assistance to countries listed by the United States as supporters of terrorism become subject to the same sanctions.

18. Other U.S. sanctions against oil-producing and transit countries are presented in appendix B.

19. *Algeria (Country Report)* (Washington, D.C.: Energy Information Administration, U.S. Department of Energy, February 1999) <http://www.eia.doe.gov/emeu/cabs/algeria.html> (September 19, 1999).

20. This would affect the economic well-being of every country in the world because of the interrelations that exist in the world energy market.

The UN embargo on Iraq, and separate U.S. embargoes on Iraq, Iran, and Libya,[21] have constrained the development of the petroleum sector in these three countries, which together possess 22 percent of the world's oil reserves (compared with Saudi Arabia's 25 percent share). If successfully developed, these oil reserves could be crucial in compensating for any major supply interruption.

- Concern about the prospect of diversion of technical expertise and fissile material from civil to military purposes constrained the choice of nuclear energy as an environmentally safe fuel option in the developing world.[22] Constraints on further use of nuclear power also exist because of its cost and public concerns about the safety of nuclear reactors, waste processing and disposal, and the transportation of nuclear material.

Utility of Conventional Weapons

In terms of lethality of weapons, proliferation of long-range ballistic missiles and WMD capabilities poses the most serious military threat to energy sources, but this is scarcely the only threat. Warfare with conventional weapons is more probable and can do significant damage to energy production and transportation routes.

- In the gulf, Iraq's military development is currently contained in terms of major arms imports, but Iraq still has substantial conventional capabilities. Iran, although weakened by its loss in the Iran-Iraq war, still has significant conventional capabilities that it can bring to a regional conflict.
- In the Caucasus, conventional arms as well as terrorism have been used to express the separatist goals of Dagestan and Chechnya.[23] Armed conflict has occurred between Azerbaijan and Armenia over Nagorno-Karabakh and could occur again if the peace process fails.

Increased Risk of International Conflict

One of the ironies of the end of the Cold War is that the risk of a global conflict has been severely reduced, but the risk of specific conflicts within and between nations has increased. The Cold War exerted certain disciplines on international behavior. Conflict was dampened because people feared that out-of-control conflicts would lead to nuclear war. As a consequence, superpowers avoided overt conflict with a client state of the opposing superpower unless the stakes were particularly high.[24]

In 1973, fear that the Soviet Union would enter the October War on behalf of its Arab clients helped lead the United States to push for a cease-fire. Conversely, the fact that Iraq would not be able to count on the support of its former Soviet patron

21. The Iran-Libya Sanctions Act refers to attempts to acquire WMD (as well as to support of terrorism) as the rationale for the sanctions.

22. One example of this is the U.S. opposition to Russian assistance to Iran for completion of construction of the Bushehr nuclear power reactor.

23. Russia has attempted to contain these separatist movements by using overwhelming military force.

gave the United States greater room for maneuver in preparing for the Gulf War of 1991.

Absent the Cold War, the potential for an outbreak of cross-border violence in instances of internal political instability has increased. Many national governments in the developing world formerly received substantial economic and military support from their Cold War patrons. Without this support they are weakened and are less able to control internal insurrections.

Implications for the Future

The geopolitical factors just described have both positive and negative implications for the achievement of the policy goals set forth here.

Adequacy of Future Energy Supply

The end of communism, the abandonment of the socialist model of political and economic relations throughout most of the world, and the engagement of formerly isolated states in the world economy make it much more likely that measures to achieve adequate worldwide energy supplies will come to fruition. Economic liberalization, privatization, and reliance on market mechanisms, not state-controlled command economies, will free up entrepreneurial forces and enlarge opportunities to develop and distribute energy supplies.

Concurrently, however, demographic pressures of population growth, urbanization, and the industrialization of the developing world will cause dramatic increases in the demand for energy.

Reliability of Future Energy Supply

The geopolitical outlook at the beginning of the twenty-first century raises questions about the future reliability of energy supply.

Forces contributing to order will be diminished and forces making for turmoil will be increased; the end of Cold War discipline will contribute to more international conflict. But the prospect of a global conflagration is substantially reduced. Conversely, some nation-states—especially in the developing world—will have to contend with increasing pressures of ethnic separatism. Nation-states and international organizations will have less control over the political agenda; increasingly they will share power with private groups of national and international activists.

In repressive regimes, for example, indigenous groups that seek to challenge the existing leadership will find their power increased when international NGOs come to their aid. NGOs can bring their communication skills and financial resources to bear on the problems at hand—human rights abuses, environmental degradation, and the absence of democracy, among others. Although the goals of such activity

---

24. One example of high stakes was the U.S. intervention on behalf of the South Korean government—under UN auspices—after South Korea was attacked by North Korea; also, U.S. troops entered Vietnam on behalf of the South Vietnamese government. In both instances the local adversary was supported by the Soviet Union.

will be laudable, political instability will increase in the short term before the benefits of political improvement can be achieved over the longer term.

In addition, the strain of demographic pressure and the various dislocations associated with economic development will make it more difficult to maintain political stability in many of the developing countries, including those that are large energy producers. Energy wealth can be used by these regimes to assuage political and economic discontent, but pressures will build over time and the income derived from energy production can be stretched only so far.

The use of terrorism and an increased availability of conventional weapons and WMD will enhance the ability of countries and separatist groups to pursue political goals by means of force. Consequently, the production and transit of energy supplies will be placed in jeopardy—at specific times and in specific locations. In most instances, such supply disruptions can be managed with little international economic damage. But if there is a supply disruption of sufficient scope and duration, worldwide energy prices will soar, causing inflation and leading to compensatory macroeconomic policies that could trigger a global recession. Changes in leaderships, however, do not necessarily lead to sustained supply interruptions. New leaders need the oil-derived income just as much, if not more, than those replaced.

Protection of the Environment

Demographic trends, coupled with the trend of economic development itself, will increase the environmental challenge posed by the continued use of fossil fuels. Technologically advanced countries will be able to pursue research and development (R&D) programs that could yield new technological solutions to various environmental problems, but without dissemination in the developing world their favorable impact will be too narrowly distributed to achieve real results globally. Will talk be matched by action?

How can this new technology, or even off-the-shelf clean-coal technology, be made available to developing countries constrained by continuing hard currency shortages? Is there a role in such technology transfer for international financial institutions that would be able to ease the financial burden? After all, actions taken in any country to reduce pollutant emissions benefit everyone.

Many of the NGOs of private activists experiencing an increase in political power have an environmental agenda. This guarantees that environmental issues will remain in the forefront of the energy debate.

CHAPTER 3

An Adequate Supply of Energy

When policymakers consider the possibility of inadequate energy supply, they usually consider the supply side of the supply and demand equation. But just as a significant decrease of supply in the energy market would increase prices and potentially cause worldwide inflation and recession, so would a substantial increase in demand if compensating supplies were not readily available.[25] Everyone assumes that energy demand will increase substantially during the coming two decades. Until now, most observers have assumed that the increased demand will be matched by a commensurate increase in supply. Planners envision several scenarios during 2000–2020 that could cause increased demand to run up against substantially tightened levels of energy supply.

- Oil markets could contract as early as 2010 if production of conventional oil peaks[26] and a dislocating transition to the use of unconventional oil is necessary. Although some believe this will happen,[27] SEI rejects the proposition that oil production will peak by 2010.
- Supplies of both oil and natural gas could be insufficient to meet increased demand if major energy-producing countries are unable to mobilize and invest in sufficient productive capacity.
- Oil and natural gas might not reach high-demand markets if transportation infrastructure in the form of oil and gas pipelines and natural gas liquefaction facilities do not receive timely and sufficient prior capital investment and cross-border transit permission.
- The energy market as a whole could tighten if environmental restrictions on the use of coal are introduced and the demand formerly satisfied by coal shifts to other sources of energy.[28]

---

25. Note that there has been a tendency to overestimate demand requirements. There can be no assurances that present-day estimates of future demand will be any more accurate than previous estimates have been.

26. See the SEI companion document, volume 2, *The Supply-Demand Outlook, 2000–2020,* for a discussion of this issue.

27. The International Energy Agency (IEA) suggests this in *World Energy Outlook* (Paris: International Energy Agency, 1998), 103–104.

28. See the SEI companion document, volume 2, *The Supply-Demand Outlook, 2000–2020,* for a presentation of this issue.

The Special Case of Asia

Asia is a region of severely limited indigenous energy resources that boasts some of the most rapidly growing economies.[29] Individual Asian countries are keenly aware of their dependence on external energy supplies as they drive toward the completion of an industrialization process that is crucially intertwined with energy use. Even though the oil market, in particular, is global in scope, several Asian countries have experienced special difficulties during periods of disruption of Middle East–supplied oil.[30]

It is important, however, to distinguish between a sudden, acute disruption in oil supplies—which the Asian countries are not well equipped to handle—and longer-term structural changes in the oil market, which some of them can accommodate.

When a sudden disruption of oil supply occurs, the higher import dependence of the Asian countries and their lack of emergency government-controlled inventories (except in Japan and South Korea) mean that they will bid spot prices up and exaggerate the temporary price spike.

But when there are structurally determined upward swings in oil prices as part of a commodity cycle, the underlying competitiveness of Asia's industrializing economies enables the Asian countries to withstand the impact of higher oil prices. Most Asian countries have rates of gross domestic product (GDP) growth that are higher than growth rates in North America and Europe. Moreover, in a tightened market and high-oil-price environment, the oil exporters can be expected to spend a portion of their additional revenues on purchases of goods and services from Asia, thereby contributing to a more favorable balance of trade for Asia.

India, whose size makes it a major factor in rising energy demand despite projected growth rates lower than the Asian "tigers," has a somewhat more vulnerable balance of payments and will be more adversely affected by increases in oil prices.

Apart from the broader world market for energy, the Asian countries are in an inherent situation of energy competition on a regional basis. Important differences among the Asian countries can already be observed in terms of population size, baseline levels of consumption, and intensity of energy use. Differences will develop in rates of continued industrialization and of increased demand for energy for electricity and transportation. Differences will also emerge in the rates at which they develop a middle class with an increased hunger for energy.

As an illustration of this point, see table 3.1 for projections of the Asian market for oil for the period 2000–2020 and present changes in consumption levels for the major countries as a proportional share of the market.

---

29. See Daniel Yergin, Dennis Eklof, and Jefferson Edwards, "Fueling Asia's Recovery," *Foreign Affairs* 77, no. 2 (March/April 1998), 34–50.

30. Kent E. Calder, *Pacific Defense: Arms, Energy, and America's Future in Asia* (New York: William Morrow, 1996), 46, describes South Korea as having had trouble competing for oil during the two oil shocks of the 1970s and during the Gulf crisis of 1990–1991. Calder states that South Korea had more trouble than Japan did during these periods because it lacked Japan's political-economic clout.

Table 3.1. Asian Oil Consumption, 2000, 2010, and 2020, as percentage of total oil imports by Asia

Countries	2000 (%)	2010 (%)	2020 (%)	Percent Change in Share, from 2000 to 2020
Japan	27	23	20	-26
Australasia	6	6	5	-17
China	23	25	27	17
India	9	12	13	44
South Korea	10	13	14	40
Other Asia	25	22	21	-16
Total	100	100	100	—

Source: Percentages calculated on the basis of data presented in "World Oil Consumption by Region, Reference Case, 1990–2020," *International Energy Outlook 1999,* (Washington, D.C.: Energy Information Administration, Department of Energy, March 1999), 145, table A4.

Note: Totals may not equal 100 because of rounding.

During 2000–2020, Japan's proportional share will decrease by 26 percent, while India's share rises by 44 percent, South Korea's rises by 40 percent and China's rises by 17 percent. The Pacific Rim countries are classified as "Other Asia" and their proportional share of the Asian oil market decreases by 16 percent.

Although China's proportional increase in the Asian oil market is expected to be less than one-half of South Korea's—South Korea has no oil of its own—it is important to remember that Asia's largest energy markets will require other fuels as well (see table 3.2).

Table 3.2. Asian Energy Consumption, 2000 and 2020, by select country and by select fuel

	Oil (MMbbl/d)			Natural Gas (Tcf)			Coal (MMst)		
Country	2000	2020	Increase %	2000	2020	Increase %	2000	2020	Increase %
China	4.6	8.8	91	1.3	9.5	631	1,682	3,540	110
India	1.9	4.1	116	1.2	5.0	317	371	536	44
Japan	5.6	6.6	18	2.1	4.4	110	135	172	27

Sources: Percentages calculated on the basis of data presented in "World Oil Consumption by Region, Reference Case, 1990–2020"; "World Natural Gas Consumption by Region, Reference Case, 1990–2020"; and "World Coal Consumption by Region, Reference Case, 1990–2020," *International Energy Outlook 1999,* tables A4, A5, A6.

If projections are true, the future belongs to natural gas in terms of growth; and, clearly, in terms of relative growth in energy consumption, the future belongs to China.

The impact that demand from China will have on world energy markets over the next 20-year period is based not only on the size of its population. China in

2000 consumes very little energy when measured on a per capita basis.[31] Chinese per capita energy consumption is only 54 percent of the world per capita average[32]—an average that is already lowered by the 2 billion people of the world who do not have access to electricity or other modern forms of energy. Stated another way, per capita consumption of energy in China is one-twentieth of that in Japan and one-thirtieth of that in the United States.[33]

China is embarked on a rapid course of industrialization and is reaching the stage at which its population will want to enjoy the lifestyle and standard of living already attained by other industrializing countries. Energy decisions that China is making now and will continue to make over the near term consequently will have a profound impact on the world energy situation during the next two decades:

- The fuel that China chooses for the government-sponsored, Chinese "people's car" will have an impact on the tightness of the world oil market; and
- The extent to which China is successful in replacing coal with natural gas will have an impact on the global environment.

It is increasingly clear that all countries are interdependent in matters of energy and the environment. Therefore, it is in the self-interest of other countries to try to ensure that China is able to meet its energy needs in ways that do not cause political tensions with its neighbors or are destructive of the efforts of other countries to take positive steps to address environmental problems.

Regional Rivalries

Future energy competition in Asia will not take place in a geopolitical vacuum. Asian tensions and enmities have historical roots, and rivalries are inherent in the geography of the area. Japan has a history of troubled relations with China, Korea, and Russia and somewhat less tense relations with the Southeast Asian countries. India and China are natural geopolitical rivals and have a history that includes border warfare.[34] China and Russia are also geopolitical rivals, as are China and Japan. A unified Korea (which could well come into being during the first 20 years of the new century) would add another strong nation to the Asian balance-of-power

31. Calder, *Pacific Defense,* 49.

32. This percentage is calculated on the basis of data presented in "World Total Energy Consumption by Region, Reference Case, 1990–2020 (Quadrillion Btu)" and "World Population by Region, Reference Case, 1990–2020," *International Energy Outlook 1999,* table A1, table A16. In 2000, China's energy consumption is 43.6 quads (quadrillion Btu) and its population is 1,231 million; for the same year, world energy consumption is 402.7 quads and world population is 6,168 million. Thus, China's per capita energy consumption is .035 quads, and the world level is .065 quads. (These are actually EIA projections for 2000. The most recent data are for 1996, but the relations are comparable: China's energy consumption in 1996 was 37.1 quads for a population of 1,220 million; world energy consumption was 375.5 quads for a population of 5,687 million. The per capita energy consumption for the world was .066 quads; the per capita energy consumption for China was .030 quads—46 percent of the world's per capita level.)

33. Allen Hammond, World Resources Institute (presentation at "Seminar on Vision 2010: Thinking about the Future," a conference sponsored by the U.S. Department of State, Meridian House, Washington, D.C., September 13, 1999).

34. In 1962, in the Himalayas.

equation. The region is home to the world's most active dispute between nuclear-armed states—India and Pakistan. Conventional border fighting occurs almost daily, and political relations are deeply troubled.

In addition, still-divided countries confront each other: China and Taiwan, and North Korea and South Korea. Each could break into overt military conflict that would have broader regional implications and would trigger implementation of U.S. security commitments to Japan, Taiwan, and South Korea.

A variety of conflicting territorial sea claims provides additional tension. China has asserted sovereignty over almost all of the South China Sea, including lanes that Japan uses to transport 70 percent of its international trade. Conflicting maritime claims between Vietnam and China have already led to violence in the Spratly Islands; and Chinese territorial claims are contested by the Philippines, Malaysia, Thailand, Brunei, Taiwan, and Vietnam in other parts of the South China Sea. Unexploited hydrocarbon resources are believed to lie underneath the disputed waters—and this raises the stakes considerably.[35]

The geopolitical landscape in which any Asian energy conflict would occur is one in which all sides are heavily armed.[36] The tensions and armaments that already exist in Asia do not mean that overt conflict is inevitable. Competition for energy among the Asian nations does not have to be expressed in military ways. But the combination of energy salience, energy competition, regional tension, and military capability add up to a situation in Asia in which care must be taken to keep energy issues from becoming unduly divisive.

The U.S. Role

U.S. protection of world sea lanes is particularly important in the Asia-Pacific region and will be increasingly so during the years ahead. This dependence will only grow with the projected increase in Asian dependence on energy exports that come from the Persian Gulf.

Meanwhile, Asia cannot assume that the U.S. electorate will continue to devote its tax dollars to protecting the energy supplies of Japan, China, India, Korea, and the countries of the Pacific Rim. This issue of burden sharing will arise as the U.S. public realizes that the United States is able to get most of its own imported energy

---

35. Recent exploration in the South China Sea has yielded disappointing results, however.

36. Large standing armies are maintained in China (2,930,000), North and South Korea (1,761,000), Russia (1,200,000), and India (1,145,000). Even Vietnam has an army of 572,000. Although the armed forces of Russia, North Korea, and Vietnam have deteriorated because of economic difficulties, they are still forces to reckon with. In addition, Russia, China, and the United States (as a protecting power) have sophisticated nuclear arsenals; India has tested nuclear weapons; and North Korea is believed to have an incipient, but hidden, nuclear capability.

Kent E. Calder, in *Pacific Defense,* pages 24–25, notes also that Russia has substantial sea and land capabilities in the area, with long-range bombers and tactical nuclear weapons deployed in Vladivostok, submarines carrying ballistic missiles (SLBMs) deployed in the Sea of Okhotsk, and a Pacific fleet that includes 65 major surface combat ships and 70 submarines, 50 of which are nuclear. Russia also has an air base located at the southern tip of Sakhalin and an air strike and air surveillance capability that surrounds Japan and all the Japanese islands.

supplies by drawing on production resources in the Western Hemisphere and the rest of the Atlantic Basin.[37]

If the United States draws down its Pacific military presence for budgetary or other reasons, it is possible although unlikely that the energy insecurity will be expressed in attempts by Asian countries to secure their own energy supplies at others' expense. China could begin to develop a blue-water navy (in part because of its concerns about defending the Taiwan Strait against U.S. naval activities on behalf of Taiwan). Japan and Korea would then be strongly motivated to develop blue-water navies of their own and a naval arms race could ensue—with attendant motivations for preemptive action to thwart the efforts of one's adversaries.

Opportunities for Regional Cooperation

Alongside Asian regional rivalries are substantial opportunities for regional cooperation in the field of energy. The need is great because of the combination of Asia's burgeoning energy demand and relative lack of indigenous energy resources. The opportunities derive from both a fortunate circumstance of geography and the national economic characteristics of the Asians themselves.

- A natural geographic complementarity exists between the oil and gas in Central Asia, Russian Siberia, and the Russian Far East and the enormous future energy demand from countries in Asia.
- The national characteristics that have made for the "Asian Miracle" make it highly likely that opportunities for energy cooperation will be exploited with success if there is political cooperation. National characteristics include economic discipline, flexibility, high savings rates, rapid absorption of technology, strong entrepreneurial networks, an embrace of international markets, and, in a few countries, an extraordinary expansion of information technology–based industry.[38]

Although the distances are great and the terrain is often difficult, plans abound for bringing oil and gas to China from Central Asia, Russian Siberia, and the Russian Far East. Some of these plans include routes that would continue to Japan and South Korea. Thus China will play an important role: it will be a primary regional energy consumer, and it could be a land bridge linking the energy production centers in Central Asia and Russia to energy markets in countries farther south and east. A similar, tremendous potential for energy cooperation exists in South Asia, with India interested in buying natural gas from Bangladesh, electricity from Pakistan, hydropower from Nepal and Bhutan, as well as liquefied natural gas (LNG) and oil from the Middle East.

China also has an important energy role as it builds its internal energy infrastructure. It does not yet have a nationwide gas transmission and distribution grid, and it will also need infrastructure for nationwide dissemination of electricity. China is not unique. Asia-Pacific Economic Cooperation (APEC) is outlining goals

37. That is, from the North Sea and West Africa as well.
38. Yergin et al., "Fueling Asia's Recovery," 35.

that will include the construction of electric power grids and gas pipelines throughout Asia—eventually with interconnections that will cross national boundaries.[39]

Before any of these plans can come to fruition, immense amounts of capital investment will have to be mobilized. Building pipelines and other infrastructure will be necessary, and this will require international cooperation and investor confidence.[40] The levels of financial investment required are daunting:

- $20 billion to finance joint Chinese-Russian projects outlined in an agreement signed in April 1996;[41]
- $45.7 billion for China's planned construction of a domestic gas infrastructure;[42] and
- $1.3 trillion for infrastructure for the production and dissemination of electricity throughout the Asia-Pacific region.[43]

To a certain extent, the mobilization of capital is largely a question of organization. China and the Asian countries have much higher savings ratios than the United States and Europe. But a high level of mutual confidence must exist before these savings will be invested in cross-border joint ventures.[44] In fact, a new level of international cooperation and investor confidence will be required.[45]

Regardless of difficulties, the record of foreign investment indicates that private-sector foreign investors will be tempted to accept the political risks in return for the growth prospects of the area. But creation of long-distance natural gas pipelines and LNG facilities, because they are so capital intensive, require a firm commitment by the parties to long-term, stable political and energy relationships that can be counted on to last.

If these conditions can be met, opportunities abound:

- LNG ventures enabling natural gas to be exported from Sakhalin, Russia, to receiving facilities that would serve the rapidly growing Chinese provinces of Guangdong, Fujian, Zhejiang, Shanghai, and Jiangsu;[46]
- Long-distance natural gas pipeline routes linking Russia to both China and Korea and even Japan, as indicated in figure 3.1 on the next page; and

---

39. Xiaojie Xu, *The Gas Dragon's Rise: Chinese Natural Gas Strategy and Import Patterns* (Houston, Tex.: James A. Baker III Institute for Public Policy of Rice University, April 1999), 31.

40. Yergin et al., "Fueling Asia's Recovery," 37.

41. Xu, *The Gas Dragon's Rise,* 15.

42. Ibid., 36.

43. Calder, *Pacific Defense,* 204.

44. The consummation of such projects would be facilitated if there existed an accepted East Asian regional investment code of the same general type as the European Energy Charter—a point made by Martha Harris of The Atlantic Council, and others.

45. Yergin et al., "Fueling Asia's Recovery," 37.

46. The demand for natural gas is high in these areas because they have limited hydropower potential and they are far away from indigenous Chinese sources of coal. Significant transportation costs have to be added to the price of coal. In addition, there is a reluctance to use coal because of the local pollution that coal-fired plants have caused. Xu, *The Gas Dragon's Rise,* 33.

Figure 3.1. Selected Gas Export Pipeline Proposals from Eastern Russia

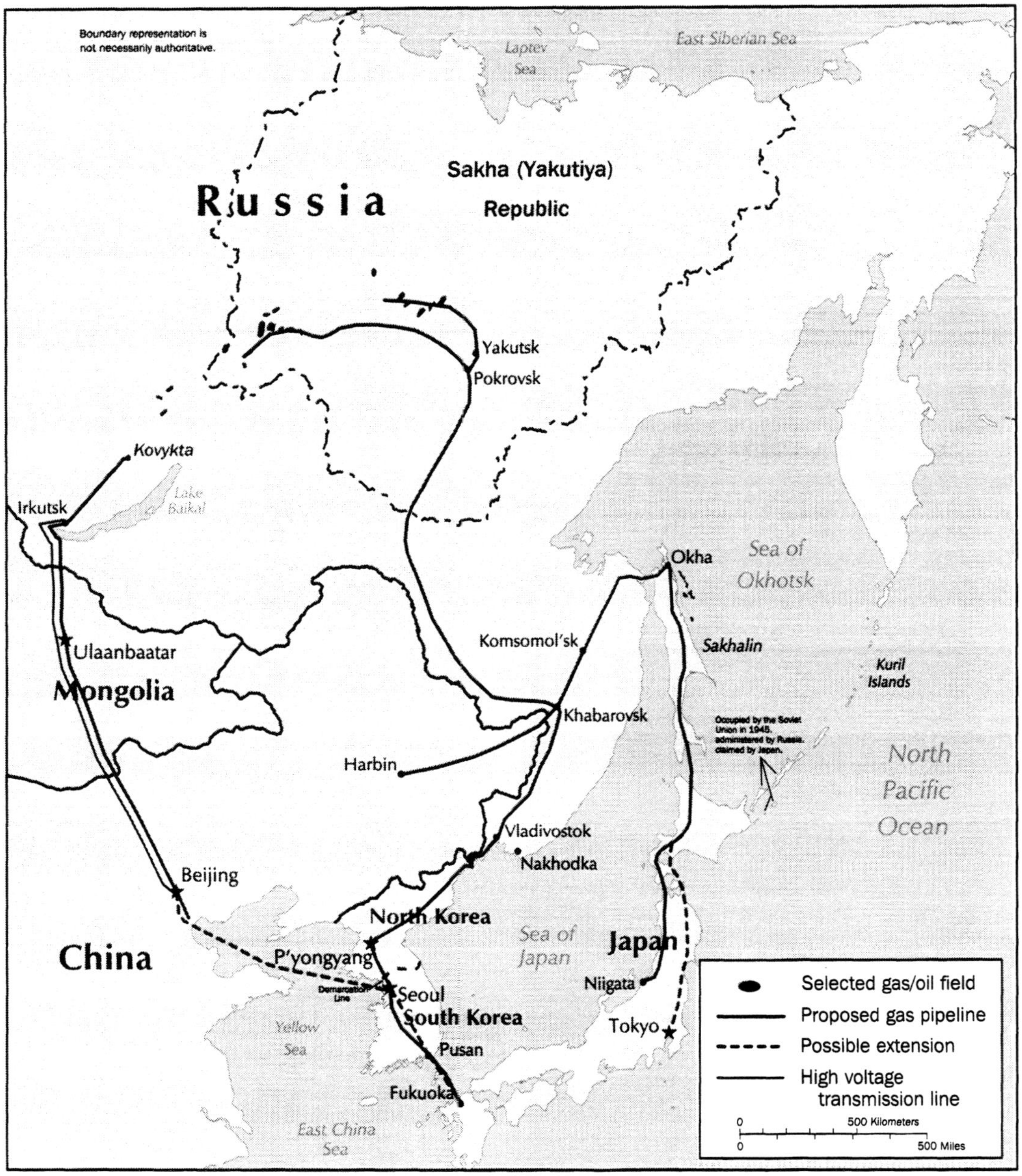

- Pipelines supplying West Siberian oil to China, either via Mongolia to Beijing or via China to Daqing.

A variety of cross-border investment relations are already at work and are likely to be reinforced during 2000–2020. Japan and South Korea are actively investing in hydrocarbon resources in Russia and Central Asia and are looking for opportunities to cooperate with China in the field of energy transportation.[47] Meanwhile, Chinese "oil majors" have concluded a substantial upstream investment arrangement in Kazakhstan,[48] made commitments to invest in oil fields in Venezuela, committed $1.2 billion to develop the Al-Ahdab oil field in southern Iraq, and reached an agreement to form a joint venture with the National Iranian Oil Company for exploring for offshore reserves in China, Iran, and elsewhere.[49]

Closer to home, China maintains a presence in oil and gas trading businesses in Singapore, Malaysia, and Thailand and supports oil and gas development in Thailand, Myanmar, and Bangladesh.[50]

These cross-border investments have been occurring on an ad hoc basis, as commercial opportunities arise. But larger infrastructure investments may need new cooperative, multilateral mechanisms[51] that would help diffuse financial and political risks and, ideally, would keep national rivalries in check. And, depending on which countries invest, cooperative geopolitical goals could also be served.

Increasing energy supply to Asia in a noncontentious manner would have global as well as regional benefits. The more secure the energy-short Asian nations perceive themselves to be, the less likely it is that they will engage in competitive political relations with key Middle East producers. They would also be less likely to engage in geopolitical rivalries for control over the East and South China Sea supply lines from the Middle East.

Cooperative joint investment ventures in major Asian infrastructure projects could help bind the investing countries together. For example, joint investment ventures between the United States and Japan in energy projects in Asia could reinforce the positive bilateral relations that already exist and make U.S.-Japanese energy competition less likely. But political rivalries could impinge upon commercial projects.

Most important, perhaps, would be the beneficial effect on the relationship between investor and recipient countries. U.S. energy investment in China, for example, could serve to defuse geopolitical rivalry between the two countries and thereby yield longer-term strategic benefits to both.

Concurrently, Chinese dependence on Russia for significant portions of its energy supplies and the construction of permanent pipeline infrastructure between them will reinforce a Russian-Chinese strategic relationship. Although this would

47. Ibid., 20.

48. The China National Petroleum Corporation (CNPC) has taken a 60 percent stake in one of Kazakhstan's oil companies, Aktyubinskmunaigaz, to help develop its oil fields. Plans, (currently stalemated) exist to bring Kazakh energy resources to China via a 1,900-mile pipeline to China's populous east coast. Yergin et al., "Fueling Asia's Recovery," 42.

49. Ibid.

50. Xu, *The Gas Dragon's Rise*, 21.

51. Calder, *Pacific Defense*, 203, 219, 222.

have the effect of defusing potential tensions between the two countries, it would also present the world with a new and stronger center of geopolitical power.

In sum, energy security pressures (in Asia) may be relieved through commercial arrangements requiring cooperation and integration. And, with the opening of energy markets to international investment, energy security will increasingly be based on market competition instead of on government edict.[52]

The Caspian and Pipeline Politics

Development of the oil and gas potential of the Caspian Sea region has become hostage to pipeline politics as both the United States and Russia seek to advance their national interests in the region. The United States has long supported the Baku-Ceyhan, Turkey, crude-oil export pipeline as well as the trans-Caspian pipeline designed to carry natural gas from Turkmenistan to Turkey. Two reasons for the unswerving U.S. support have been the denial of Iran as a pipeline transit country and the minimization of Russia in the movement of oil out of the region. But U.S. policy appears to have been founded on political considerations more than on commercial realities and thus has not gained the necessary wide support from foreign investors. Pipelines are not built to support political interests; pipelines will be built when sufficient oil is available to justify construction or when purchase contracts have been signed for natural gas, and not before.

According to the Azerbaijan International Operating Company (AIOC) that is developing production in the Azeri sector of the Caspian Sea, the Baku-Ceyhan pipeline has yet to move off the drawing board because known oil reserves—4 billion barrels—fall short of reserves required—6 billion barrels—to justify its construction in 2000. Moreover, some AIOC member-companies have indicated they may prefer to export their equity crude oil to foreign markets by routes other than Baku-Ceyhan.[53] Baku-Ceyhan is a pipeline for the indeterminate future.

The future of the trans-Caspian gas export pipeline is even dimmer, challenged in part by the discovery of natural gas at Shah Deniz, also in the Azeri sector of the Caspian Sea. Plans are being developed to deliver Shah Deniz gas to markets in Turkey by the beginning of the 2002–2003 heating season, a timetable that the trans-Caspian line cannot meet.

Russia has negotiated to purchase substantial volumes of natural gas from Turkmenistan, beginning in 2000. For Russia, importing Turkmen natural gas serves a dual purpose: supporting continued Russian natural gas sales to Europe and supplying hard currency and, at the same time, sharply delimiting Turkmenistan's export capability and further negating prospects for the trans-Caspian line.

Oil and natural gas development and related pipeline activity in the Caspian region, including Central Asia, cannot be isolated from politics. Much is at stake, and both Russia and the United States recognize that. It is a struggle for influence, for political linkage, and it is taking place in Russia's back yard.

---

52. Yergin et al., "Fueling Asia's Recovery," 46, 48.
53. David Stern, "Baku Pipeline Runs into Hitch," *Financial Times,* June 11, 2000.

Russia of course is doing what it can to present itself as a reliable and logical transit country and, in that regard, is doing quite well. Considerable oil volumes have been flowing northward out of Kazakhstan by pipeline to Samara in Russia. Construction of the Tengiz-Novorossiisk oil export pipeline—known as the CPC–Caspian Pipeline Consortium—appears on schedule to meet its mid-2001 completion date. The larger portion of this pipeline passes through Russia. Construction of Blue Stream, a gas pipeline to carry Russian natural gas beneath the waters of the Black Sea to Turkey, is under way. In 2000, the Baku-Novorossiisk oil export pipeline is again operational (a bypass was built around Chechnya), and a short link is in place between this pipeline and the west Caspian port of Makhachkala to facilitate oil exports from Kazakhstan and Turkmenistan. Clearly, at midyear 2000, Russia appears to hold the advantage in the great game of pipeline politics.

CHAPTER 4

A Reliable Supply of Energy

A geopolitical crisis could cause an energy supply disruption of sufficient magnitude to result in adverse worldwide macroeconomic effects.[54] In theory, international conflict and domestic turmoil could interfere with the production and transportation of coal, natural gas, and oil; but, because of oil's prominence in international trade and the reliance of oil consumers on a small number of key oil suppliers, this chapter concentrates on the effect on oil.

International conflict is more likely to occur during the decades of 2000–2020 because Cold War disciplines have been removed. International power is newly configured and has been described as "uni-multipolar,"[55] with the United States enjoying power preeminence in a context that includes several strong regional power centers. But, although the United States is characterized as "the world's only superpower," its diplomatic resources and military forces cannot possibly dampen or respond to every flashpoint of international conflict. Nor can the UN and regional security institutions fill the breach.

Meanwhile, weapons proliferation and the availability of terrorism as an instrument of state policy have increased the power of formerly weak states.

Increased political instability is likely. The power of some national governments and international treaty-based organizations has been diluted by the more prominent role played by private international organizations of activists. By addressing issues of human rights, democratization, and the environment that arise within states' domestic political boundaries, these private organizations have been able to increase the power of politically weak domestic groups in confrontations with their own national governments.

In addition, demographic pressures and divisive ethnic loyalties are increasing turmoil in the developing world. Ethnic separatists often receive funding from more prosperous compatriots in other countries. They take advantage of the widespread international market for both legal and illegal conventional armaments. In addition, instruments of terrorism are available to separatist groups that often use

54. Because oil is a fungible commodity and is traded in an atomized global market, any substantial interruption in oil supply will have a ripple effect that will be felt quickly by the market's many participants. The global economic damage that can be caused by any given supply interruption will be a function of several factors: (1) the magnitude and duration of the interruption; (2) the ability and willingness of other suppliers to make up the shortfall; (3) the adequacy and global diversity of strategic reserves and scarcity-sharing mechanisms; (4) the ability of the market to operate in ways that discourage hoarding and panic purchases; and (5) the response of oil importers in terms of macroeconomic adjustment or accommodation policies (for example, if inflation is low, oil price increases, especially if temporary, can be accommodated).

55. This term was introduced by Samuel Huntington in *The Clash of Civilizations* (New York: Simon and Schuster, 1996).

them to redress the inherent power imbalance they confront when they oppose established national governments.

Supply Interruptions Caused by International Conflict

Regional Conflicts

The last significant oil supply interruption was in 1990–1991, but it is unreasonable to assume that the world will be as fortunate during the next two decades. The number of major oil suppliers remains small, and the concentration of world oil supply among them will increase because of the expected increase in the proportion of global production contributed by the high-reserve, low-cost producers of the Persian Gulf.

The possibility of regional conflict in the Middle East and gulf area can never be discounted. A conflict that involved the use of WMD could be devastating. The gulf is the energy-producing area of the world that is most vulnerable to the prospect of a conventional conflict severe enough to interrupt its contribution of oil and gas supplies to the world market. Even a conventional war could completely halt the exports of one or more gulf producers, just as the embargo that followed Iraq's invasion of Kuwait halted all oil exports from Iraq until the outcome of the Gulf War reversed the situation. Iraq already has a record of using force in the gulf in support of its energy policy and regional ambitions. It may well seek to repeat its 1990 invasion of Kuwait and has sufficient ground forces to break out from Basra and overrun Kuwait quickly—unless an advance warning permitted strong reinforcement of Kuwait's existing defense positions.

A lower—but still significant—possibility exists that regional conflict in Asia could break out because of long-standing regional enmities, geopolitical rivalries, and competition for scarce oil supplies by the rapidly developing countries there. Even if only of a harassing nature, such conflict could impair the free flow of oil to consuming countries in the region, thereby setting up a competitive inventory-building dynamic among them that would distort global oil market relationships and exert a downward pull on total global oil supply.

Cross-Border Subversion

In several situations—especially in the gulf—adversaries can heighten and exploit potential sources of internal instability. The level of support given by one major oil producing country to the dissident elements in a competing producing country could become an additional way of pursuing internal OPEC policy divisions (over production quotas, for example). Pursuit of strategies of cross-border subversion against a competing producer could have an important harassing effect, although not the power to overthrow, and would be a hard-to-ignore way of signaling displeasure. The impact of attempts to contribute to the destabilization of other countries is twofold: it can result in actual supply interruptions, or it can put in place U.S. sanctions and deter investment in needed production and pipeline facilities, thereby constraining future adequacy of energy supply (see page 22).

- Iran could strengthen its current efforts to support dissident groups in Iraq, particularly when internal divisions and internal unrest in Iraq become serious. Iran could weigh in covertly to provoke a new Shiʿite uprising in the South of Iraq or, at some risk to itself in view of its own Kurdish problem, give Iraq's Kurdish separatists enough covert support to cause turmoil and perhaps secure independence in areas of northern Iraq where major oil fields are located. In addition, Iran could seek to exploit the fracture lines and sources of political unrest within and between the southern gulf states. This is particularly true with respect to Saudi Arabia, where Iran's coreligionist Shiʿites represent an ethnic majority in the eastern province of Saudi Arabia where major oil fields are located;[56]
- Iraq could continue to strengthen its support of dissident elements in Iran and could seek means of exploiting any major sign of internal unrest in competing oil producer Saudi Arabia; and
- Institutions or individuals, without the knowledge or support of their national governments, may support terrorism and subversion. A case in point is Usama bin Laden, an Arab born in Saudi Arabia who is accused on masterminding—and financially supporting—attacks on Western, particularly U.S., facilities.

Similar cross-border subversive activities could be pursued in the Caspian, particularly in support of oil export routes that favor particular countries—the pipeline wars. Elements in Russia[57] have reportedly engaged in efforts to destabilize Georgia—which is already a transit country for the Western early oil route for exports from Azerbaijan and is slated to be a transit country for the proposed East-West main export pipeline from Baku to Ceyhan. The still unresolved conflict between Azerbaijan and Armenia over the Nagorno-Karabakh enclave provides another motive and opportunity for mutual cross-border subversion that could jeopardize East-West pipeline routes. And, because the proposed Baku-Ceyhan main export pipeline would cross a major Kurdish area in Turkey, opportunities to engage in subversion there would also present themselves, albeit in a situation of strong Turkish internal security.

Supply Interruptions at Choke Points

The United States and other countries of the world have long been conscious of the risks posed by energy choke points.[58] The Strait of Hormuz, for example, is a choke point through which an estimated 30 percent of world oil exports will travel during 2000, rising to 38 percent of world oil exports by 2020. The Strait of Hormuz will

56. Iran could also seek to reinforce Shiʿa dissension in Bahrain, with the possibility of regional repercussions, and could act in ways that would exacerbate any future confrontations between Bahrain and Qatar or Saudi Arabia and Yemen.

57. It is not clear whether various attempts to increase turmoil in Georgia—by trying to assassinate President Eduard Shevardnadze, for example—were sanctioned by the Russian government or were rogue operations conducted by elements of the former Soviet security apparatus.

58. See appendix A for a list of the primary sea-lane energy transit choke points.

increasingly be the key point of passage for oil exports traveling to Asia. The Strait of Malacca is a also choke point through which gulf oil bound for Asia passes. There are incipient choke points as well.[59]

The proliferation of long-range naval strike aircraft, antiship missiles, smart mines, submarines, and guided missile ships throughout the world is greatly extending the range at which military strikes can be launched against targets associated with the movement of energy supplies. This trend is likely to continue during 2000–2020. Meanwhile, the transport of oil and natural gas via long-distance pipelines and tanker will increase because of the expected significant growth in worldwide energy consumption, particularly in Asia. Therefore, the supply security risks related to transit by water and pipeline will increase significantly during the first decade of the twenty-first century.

Over the next 20 years, energy markets will continue to grow, integrate, and electrify. Secure infrastructure will be required so that energy supplies can continue to move reliably from energy producing sources and regions to increasingly geographically dispersed end-use markets. Although physical security of oil and natural gas pipelines, terminals, and tankers will be important, the emerging energy infrastructure will be based on and controlled by the flow of electricity and sophisticated electronic information systems. If electricity fails in one small area, it could cascade into a larger regional blackout. Such a blackout could affect telecommunications and the operations of pipelines and terminals as well as operations of refineries and natural gas processing plants.

These increasingly sophisticated electronic systems frequently are both interconnected and computer dependent. Such electronic sophistication adds efficiency and reduces operating costs and, to a certain extent, also increases the ability to repair damaged facilities quickly. Such systems are also vulnerable to electronic viruses that terrorists, hackers, or even disgruntled employees could deliberately use to bring whole energy systems down.[60]

Supply Interruptions Caused by Political Instability

Theoretically, severe political instability could cause disruption of all categories of energy supply.[61] But, again, oil is most important because it will continue to be the predominant source of energy used during the next two decades. Also, oil con-

---

59. Although not listed as such, over time Baku in Azerbaijan could well become a significant choke point if oil is piped across the Caspian Sea from Kazakhstan and Turkmenistan and then through Baku for further transit west through the yet-to-be-constructed main export pipeline that would reach the Mediterranean at the Turkish port of Ceyhan. Natural gas from Turkmenistan might parallel the proposed oil pipeline.

60. Disaffected people could also introduce into the system sufficient misinformation for system operators to believe there was a genuine malfunction. System operators would themselves then bring the energy systems down.

61. In some countries of Latin America, for example, electric power transportation pylons have been subjected to sabotage as an expression of opposition to privatization of the national electric power system.

sumption is more dependent on international trade than is consumption of either natural gas or coal.

With only two major exceptions—Norway and the United Kingdom—almost every current major oil-exporting country exhibits characteristics that could make it vulnerable to internal political instability severe enough to cause its oil production to be diminished substantially. The potential impact of an oil supply disruption caused by internal political instability in one or more of the major oil exporting countries can be appreciated by examining the current situation of worldwide oil dependency.[62] The Energy Information Administration (EIA) has gathered data describing the level of net exports for the 13 oil-producing countries able to export more than 1 MMbbl/d in 1999 (see table 4.1).

Table 4.1. Top Petroleum Net Exporters, 1999 (MMbbl/d)

Rank	Country[a]	Production[b]	Consumption[c]	Net Exports[d]
1	Saudi Arabia	8.5	1.3	7.2
2	Russia	6.3	2.3	3.9
3	Norway	3.1	0.2	2.9
4	Venezuela	3.0	0.5	2.5
5	Iran	3.6	1.2	2.4
6	Iraq	2.5	0.5	2.1
7	United Arab Emirates	2.3	0.3	2.0
8	Kuwait	2.0	0.2	1.8
9	Nigeria	2.1	0.3	1.8
10	Mexico	3.4	2.0	1.4
11	United Kingdom	3.0	1.7	1.2
12	Libya	1.4	0.2	1.2
13	Algeria	1.4	0.2	1.1

Source: "Top Petroleum Net Exporters, 1999" <www.eia.doe.gov/emeu/security/topexp.html> (accessed March 16, 2000).

Notes:
a In 1998 these 13 countries accounted for slightly more than 81 percent of the total world oil exports.
b Includes crude oil, lease condensate, natural gas liquids, other hydrocarbons and alcohol, and refinery gain.
c Estimated using average oil consumption growth rates from recent years and adjusting according to analysts' judgments.
d Calculated by subtracting estimated consumption from production; columns may not add across owing to independent rounding.

Two previous supply disruptions can be used as templates against which to evaluate these indicators:

- Iran in 1979–1980. Before the supply interruptions caused by the lead-up to the Iranian revolution and its aftermath, Iranian net oil exports provided close to 15 percent of world oil exports. By the end of 1979, however, exports had fallen

62. Although both the EIA and the International Energy Administration (IEA) make select country-by-country projections of future oil production capacity for the period 2000–2020, they do not make similar projections for oil production or for oil exports.

to half that level and accounted for only 7.3 percent[63] of the world oil export market. By the end of 1980,[64] exports had fallen to an oil export market share of only 3.4 percent, between one-fifth and one-quarter of their former level.

- Iraq and Kuwait in 1990. Before Iraq invaded Kuwait, the two countries together contributed 14.7 percent of world oil exports.[65] After the invasion, exports from Kuwait ceased and exports from Iraq were curtailed by sanctions. During the three-month period following the invasion, close to 4.5 MMbbl/d of oil exports had been withdrawn from the market.[66] For 1990 as a whole, net oil exports from Iraq and Kuwait dropped to a 9.1 percent share of the world oil export market, only three-fifths of the contribution by the two countries before the invasion.

In an effort to mitigate the economic damage associated with oil supply interruptions, the International Energy Agency (IEA) was established in 1974 with a membership composed of most OECD countries. One of the IEA's core functions was the development of an emergency sharing system (ESS) that was designed to soften the blow of supply disruptions that exceeded 7 percent of IEA members' oil supplies by using government-controlled oil stocks. Recognizing the challenging structure of the oil industry and changing oil market dynamics, the IEA in the 1980s developed a coordinated early response mechanism (CERM) designed for more flexible responses to disruptions that were less than the 7 percent threshold. CERM was activated during the 1991 response to Iraq's invasion of Kuwait.

Several factors affect the impact of any particular supply interruption in the world market. The most important is the ability and willingness of other producers to make up the shortfall by increasing their own supplies to the market. The tightness of the market at the time of the disruption, the level of world inventories being held, the prevailing market psychology with respect to hoarding and panic purchases, and even the season of the year play their parts. In addition, world macroeconomic conditions at the time of the disruption help determine the strength of demand and the price elasticities that come into play.

Even though each supply disruption is unique, it is obvious that the impact on the world oil market of supply curtailment by a particular major exporter or combination of exporters will be a function of the share of the market the exporters represent. With the 1979–1980 example of Iran and the 1990 example of Iraq-Kuwait, it seems that a 15 percent share of oil moving in the international market before the disruption would certainly constitute a threshold at or beyond which the market and the world economy would be endangered if an interruption did take place. That 15 percent share, if applied to the top 13 net oil exporters listed in table

63. Annualized data for 1979.

64. Annualized data for 1980.

65. Annualized data for 1989.

66. The size of this 1990 drawdown of oil supply in the world market is almost identical to that withdrawn from the international oil market as a result of the decision in 1999 by OPEC and key non-OPEC producers to reduce their output. The total size of the oil market was smaller in 1990, however, and the Kuwait–Iraq shortfall of 1990 had a greater impact. Both times prices increased. Oil prices more than doubled after the 1999 cutbacks.

4.1, translates into a volume of oil not much more than the 4.3 MMbbl/d cut in exports imposed by oil exporters in March 2000, a cut that almost tripled the price per barrel within a comparatively short period of time.

This does not mean that all exports of a major oil market contributor would have to cease before the level of its previous market share would be relevant. Neither Iran nor Iraq ceased its oil exports altogether. The issue is one of risk and uncertainty. Perceptions held by the other players in the market, especially those trading in "paper" barrels, are crucial in determining the outcome. As soon as export levels are affected by a cross-border conflict or by serious political instability in a major exporting country, it is prudent to consider the possibility that the conditions precipitating the export curtailment will worsen, that any military conflict will increase in scope and destructive force, or that civil conflict will intensify and spread.[67]

Such prudence leads oil importers to hold off drawing on their strategic petroleum reserves in case the oil supply crisis worsens. After Iraq's invasion of Kuwait, for example, the initial major fear was that the conflict would spread to Saudi Arabia. When it became clear that the Saudis were not going to be attacked, calm began to return to the market and prices began to moderate. It can be assumed, therefore, that the very fact of cross-border military action or severe internal political turmoil involving one or more countries with significant net export market share will be enough to disrupt the international oil export market.

With these considerations in mind, market share data for the principal net oil exporters give little cause for comfort. The Saudi Arabian share of world oil trade is well above the 15 percent threshold, and its market share is expected to increase substantially during 2000–2020. Although no other single country exceeds the 15 percent threshold, turmoil in Iran or Venezuela could easily occur concurrently with long-term chaos and export curtailment in Russia, and the combined market share would clearly exceed the threshold.

If the threshold is lowered to a 10 percent share of the international oil export market, a variety of combinations whereby simultaneous distress in two or more of the major net oil exporting countries would be reason for concern.[68] Several combinations of pairs of countries would exceed a 10 percent market share threshold. The assumption is that simultaneous turmoil in two or more of these countries would be a result of random events that occurred coincidentally. Other simultaneous

---

67. Paradoxically, this kind of uncertainty may discourage importing countries from taking early action to draw upon their strategic reserves or the IEA from invoking its sharing mechanism at the outset of an oil supply disruption because, in each case, there would be a fear that taking action would prove to have been premature should conditions of even greater scarcity occur later in the conflict.

68. Wilfred Kohl, professor and director of the Energy and Environment Program at the Johns Hopkins School of Advanced International Studies (SAIS), points out that a shortfall of only 5 percent of world production would register immediate major price increases on the futures and spot markets. Note that Kohl is referring to worldwide production, a category of analysis much broader than the category of net oil exports used to establish the threshold used here. His analysis therefore serves to reinforce the argument that small interruptions of supply can result in substantial increases in price.

export curtailments could occur as a result of cross-border military operations affecting two or more of the countries of the gulf region.

Even a cursory glance at the list of 13 leading petroleum net exporters reveals that most have either experienced noteworthy internal stress in the past or contain incipient sources of stress that could flare up in the future. With the exceptions of Norway and the United Kingdom, it is not difficult to conceive of future situations in most of the other 11 countries that could lead to internal strife at a level sufficient to cause a reduction of oil exports. Chapter 5 provides further discussion of internal stresses in several important oil-exporting countries.

CHAPTER 5

Major Oil-Exporting Countries

Unfortunately pessimism prevails when longtime observers consider the outlook for stable political systems in many oil-exporting countries. And, in the United States, discussions of Middle East politics and flows of oil out of the Persian Gulf rarely touch on the implications of U.S. long-term support of Israel. U.S.-Israeli policy does indeed lurk behind gulf oil exports, but few posit this support as a source of potential disruption of normal oil flows out of the region.

Egypt and Syria attacked Israel in early October 1973, in the midst of negotiations between OPEC and the oil companies on the matter of price. Negotiations had not proved successful, and the war was going badly for Israel, which turned to the United States for material aid. Arab oil ministers then took up oil as a weapon and imposed an embargo—eventually embracing the United States, the Netherlands, Portugal, Rhodesia, and South Africa—as well as staged production cuts of 5 percent per month. Saudi Arabia, reacting to the U.S. assistance given to Israel, cut off all oil deliveries to the United States. Iraq did not join in the embargo but instead increased production to take advantage of the high prices.

It seems reasonable to consider again the prospect of a future embargo by an Arab oil exporter that takes exception to the continuing U.S. support of Israel.

The following geopolitical assessments stress prospects for instability and for interference in oil exports, but only as part of a larger effort to alert world policymakers to the fragility of reliable and timely oil flows. Complacency can never be justified. Generally speaking, the United States deals with energy matters in domestic terms, not international. The globalization of energy, especially oil, argues that this approach should be reversed.

Saudi Arabia

Any geopolitical analysis of world oil inevitably begins with a survey of Saudi Arabia and includes a scenario under which its oil exports are reduced or even eliminated. That is so because of the past, current, and future dominance of Saudi Arabia as a supplier of oil to the world market. Its leadership role remains unchallenged. At the same time, its dominance causes uneasiness on the part of consumers who clearly understand that no other supplier can replace Saudi Arabia should the need arise.

There are several sources of potential instability in Saudi Arabia that could intensify during 2000–2020 to cause a major interruption in Saudi oil exports.

Structural Economic Problems

Saudi Arabia's structural economic problems are caused primarily by its burgeoning population growth and an ingrained system of social welfare payments that the government can no longer afford. When oil export revenues remain relatively static, public-sector employment expands more slowly because of budget constraints. Private-sector enterprise is handicapped by the double disadvantages of wage distortion from the public and petroleum sectors and dependence on government contracts and licensing in an economy in which administrative processes are not open and transparent.

Widespread unemployment exists among young people, who are educated in a system that is not necessarily responsive to the demands of a technically sophisticated modern economy. Generous social welfare benefits lead to chronic budget deficits. Meanwhile, the combination of high population growth and low oil prices caused the entire country to experience a sharp decline in per capita income. GDP per capita fell from a level of $17,000 during the early 1980s to only $7,000 by the time of the low oil prices of 1997 and early 1998.[69] Although the government is committed in principle to achieving the economic goal of diversification with less dependence on oil income, such diversification has been difficult to bring about; this is a situation mirrored in most, if not all, developing oil-exporting countries.

Ethnic and Social Tensions

Ethnic and social tensions are likely to be exacerbated in any future situations of economic deprivation. The most prominent ethnic cleavage is also the religious difference between Sunni and Shi'a Muslims. Tension between traditionalists and modernizers is apparent. There is also a deficit of democratic institutions. A new generation in Saudi Arabia (and elsewhere in the gulf) may be increasingly dissatisfied with arrangements whereby much of the available stream of income is allocated by the ruling family. The implicit convention that the ruling family will provide social services, jobs, and freedom from taxes for the entire native population is coming under strain as the Saudi population grows (at one of the most rapid growth rates in the world).

Problems of Royal Succession

Because it is ruled by a royal family, the path of succession that the family chooses is crucial to Saudi stability. Although major succession crises have been avoided in the past and the plan to replace King Fahd with Prince Abdullah appears to be moving smoothly, there can be no guarantee that a succession crisis will be avoided in the future. Much of what happens within the Saudi royal family and within the Saudi religious establishment, the ulema, is opaque to Western observers.

The most likely scenario for political change in Saudi Arabia that would threaten continuation of Saudi oil exports might not be the classic revolt caused by

69. The most recent available per capita income statistics are for 1998. With the oil price increases that began in the spring of 1999, the Saudi per capita income figure will rise accordingly, as will the per capita income of the other oil exporters.

ethnic or class differences or in response to economic deprivation. The Shi'a have grievances, to be sure, but Saudi security forces have the situation in the eastern province well under control; and it might be later than 2020 before the modernizers in Saudi Arabia have enough power to challenge the system.

Russia

Russia lost its world leadership in oil production as its output declined from its 1988 peak of 11.5 MMbbl/d to approximately 6.1 MMbbl/d in 1999. This decline was caused by neglect and poor oil field management, followed by economic dislocation that occurred with the end of communism in 1989 and the breakup of the Soviet Union in 1991. In prior years, there had been insufficient investment in Russian oil production facilities and regular physical maintenance suffered. It is unlikely that Russian oil will ever return to its earlier level of prominence.

After a prolonged decline, the Russian economy improved from 1998 to 1999, with a GDP increase of 3.2 percent and industrial growth of 8.1 percent. Much of this rebound was attributable to the import substitution effect after the ruble devaluation of August 1998, increased revenues from the higher world oil price, and some restructuring. Oil output is expected to continue to grow, but at a modest rate of 2–3 percent a year for the period 2000–2002. Domestic consumption of fuel is expected to rise gradually, permitting Russia to maintain its position as a leading exporter of oil and natural gas.

Circumstances necessary to attract significant foreign investment to rehabilitate and expand the oil sector do not yet exist. Although the country is beginning to recover from the 1998 devaluation of the ruble, the rule of law necessary to ensure predictable business relationships is unformed. The economic atmosphere of the country is tainted and made more complicated by mafialike forms of extortion and other lawless activity. Important business transactions often depend on insider relationships with government officials and leading banks that are neither transparent nor particularly reliable.

Growing feelings of nationalism may preclude the Duma from passing all the legislation needed for production sharing agreements (PSAs) to go forward. Without foreign investment through PSAs, oil production and export levels could be difficult to sustain and not likely to increase by much. Vladimir Putin, elected president on March 26, 2000, has repeatedly stressed the need for more foreign direct investment (FDI) and PSAs. A more compliant Duma will probably support him in this.

In his action program, published on the Internet on December 28, 1999, Putin stressed the low level of the Russian economy. Russian GDP, he stated, is only one-tenth of that of the U.S. GDP. He blamed this on declining domestic investment—it is only one-fifth of the 1991 level—and on the wretchedly low amount of FDI. Russia has attracted only $12 billion in cumulative new FDI since 1991, compared with more than $300 billion in China. Putin's action program and other statements indicate that he will give priority to attracting FDI and expediting PSAs and other provisions to make Russia more attractive and more receptive to foreign trade and investment.

Circumstances in Russia could deteriorate further, however, and in ways that could eventually disrupt Russian oil exports and natural gas deliveries to Europe.

Political Factors

At a political level, the power vacuum caused by the disintegration of the state apparatus formerly run by the Soviet Union's Communist Party has not been filled—at least not in ways that are politically functional. Policy stalemate occurs because of the continuing division between various groups of reformers and the conservative former Communists.

Economic Factors

At an economic level, the deprivations and uncertainties that many Russian people have had to endure provide a painful contrast with the modest—yet reliable—economic and social safety net they enjoyed during the Soviet period. These discomforts are exacerbated by the additional contrast between their lot and the opulent lifestyle enjoyed by the entrepreneurs who have been able to flourish under the new system and the criminal elements who have been able to exploit it. Economic grievances, when added to an understandable longing among many for renewed great-power status, could easily combine to provide support for a nationalist-authoritarian political movement that could contribute to countrywide instability.

Dissatisfaction within the Military

Many of the armed forces are not being paid and are engaged in nonmilitary activities, such as local farming, to support themselves and their families. Of all of the segments of society, they probably miss the country's former international strength the most, and they are said to resent domestic missions in which they are used as a national police force,[70] instead of as a regular army. Military units loyal to President Boris Yeltsin played a crucial role in 1993 when his power was contested by the Duma and he was almost ousted from his position. It is not clear, however, how the military would respond in a similar future situation.

Weakening Power of the Central Government

Significant forces are at work at the regional level that have the potential of weakening and fragmenting the political power of the central government. Three-quarters of the Russian republics already have their own separate constitutions and many of them claim at least some sovereignty.[71] The traditional ability of the central government to collect taxes effectively and provide expected services in return has been weakened to such an extent that power is devolving to the republics themselves and

---

70. Steven R. David, "Saving America from the Coming Civil Wars," *Foreign Affairs* 78, no. 1 (January/February 1999): 113.

71. Ibid., 114.

to even smaller units within them. Ties between local military units and local governments have been strengthened because the units are becoming dependent on these governments for food, housing, and wages.[72] If conflict breaks out between regions, regional military loyalties could be called into play, thereby worsening the situation and increasing the danger. President Putin, to the extent he can, is taking steps to return power to Moscow.

Ethnic Conflict

Russian chauvinism and the legacy of totalitarian and ethnically insensitive Soviet policies toward national groups have reinforced trends toward ethnic separatism. The Russian military effort to overwhelm ethnic separatists in Chechnya has been popular with the Russian public, but an unsuccessful outcome or widespread use of terrorism within Russia by Chechnya's supporters could cause the conflict to weaken the Russian regime.

Russia is the world's second largest exporter of oil and, even with the expected increases of oil production in the Middle East during 2000–2020, Russia will remain an important player in the world oil market. If the sources of instability outlined above lead widespread strikes of oil workers or general social and economic chaos, a significant disruption of expected Russian oil exports could occur some time during 2000–2020.

The exact characteristics of such a disruption cannot be predicted, but it is possible that the crucial Black Sea export port of Novorossiisk could be shut down or that the Druzbha crude oil pipeline, which moves more than 1 MMbbl/d into Central Europe, could be closed. Russia is also the world's largest trader of natural gas and is a key supplier to European markets. Most of the natural gas in 2000 is exported through Ukraine and thus is doubly jeopardized, vulnerable to internal events in both Russia and Ukraine.

Russia recognizes the vulnerability attached to having most of its gas exports to Europe transiting Ukraine and is taking steps to reduce, if not eliminate, that vulnerability. Russia currently is the largest supplier of natural gas to Turkey; this carries considerable political and economic implications. Construction is under way of an undersea gas pipeline from Russia to Turkey as Russia seeks to protect and expand its market share in that country.

Venezuela

There continues to be more than a moderate risk of increased turmoil in Venezuela,[73] certainly over the next few years and perhaps beyond. Such turmoil in a country that is the world's fourth largest oil exporter and sixth largest producer could present serious problems for the world market. Venezuela is undergoing a crisis of political and institutional legitimacy, with ensuing socioeconomic ten-

---

72. Ibid., 113.

73. This analysis has been contributed by Lowell R. Fleischer, senior associate, Latin American affairs, CSIS.

sions. The electoral mandate of President Hugo Chavez Frias, elected in December 1998, was impressive, but the political pedigree of the leadership makes the future of democratic governance uncertain.

President Chavez's most ambitious act so far has been the establishment of a constituent assembly that drafted a new constitution that was submitted to a referendum on December 15, 1999, and approved. Although the supreme court of Venezuela has ruled that the assembly's only job is to write a new constitution, at Chavez's urging it is rapidly taking over the functions of government. The assembly has already acted to reorganize the court system and remove some of the sitting judges. It has moved to usurp the authority of the legislature as well.

President Chavez now has a free hand and with the assistance of the assembly—which is completely dominated by his followers—he seems to be concentrating power in the presidency. Because the new constitution lengthens the term of a president to six years and gives the president the right of self-succession, President Chavez could be in power for the foreseeable future. Organized opposition has simply collapsed. Some combination of constitutional crisis, unmet public expectations, cross-border incursions from Colombia, or a collapse of the economy would only play into his hands and reinforce his authoritarian inclinations.

The current absence of credible institutions of civil society and the existence of limited economic options will likely prolong uncertainty about the political and economic evolution of the country, the results of which may not be realized for years to come. Venezuela, in 2000, is in the midst of the worst recession in its history. A political and economic crisis in Venezuela leading to violent unrest, a collapse of the exchange rate, and default on its international debt cannot be ruled out. These factors—coupled with Chavez's attempts to micromanage PDVSA, the state oil company—could combine to cause a disruption in Venezuela's oil production and exports or, at least, minimize new investments, placing a limit on the country's export capacity. In short, Venezuela may be in for a long period of instability, the outcome of which is difficult to predict.

Iran

Iran is an example of the ability of an internal revolution to disrupt the oil exports of a country, even if that is not the revolutionaries' intent. Still important in 2000 is that, in a sense, the revolution has not yet concluded.

An Unfinished Revolution

Tension between supporters of President Ali Mohammad Khatami's civil reform agenda and supporters of a harder clerical line continues. President Khatami has received overwhelming support in his own and subsequent elections, but key foreign policy, defense, and internal security decisions are still made by the supreme leader, Ayatollah Ali Hoseini-Khamenei, in conjunction with the Council of Guardians. Although the elections themselves are free, candidates are still screened by the clerical establishment. Considerable internal instability continues. President Khatami has deliberately undertaken measures to liberalize civil society, including establishing guarantees for a free press. Yet newspapers have been closed when clerical authorities consider them threatening to the established order.

Demographic Problems

Iran faces many of the demographic problems that confront other countries in the region. Half of Iran's population is under the age of 20[74] and, because the right to vote is accorded to people in Iran at the age of 15, the attitudes held by Iran's young people carry considerable political weight. President Khatami's major sources of support for his civil reform program are young people and women whose economic and social aspirations will be thwarted in the absence of reform. The attitudes of these two interest groups will become increasingly important during the 2000–2020 period.

One of the most important aspects of Iran's current demographic situation is that no one under 25 years old in Iran has any direct memory of the Iranian revolution or the shah's regime that precipitated it. Consequently, exhortations by the clerical leadership to defend the purposes of the revolution are not likely to resonate strongly, and the attractions of a lifestyle that permit more individual freedom are likely to lead to pressure for further reform.

In contrast to current reformers in Iran who are self-consciously trying to work within the current system, the next political generation in Iran may seek a complete change of regime.

Failure to Implement Timely Economic Reforms

Perhaps the most important source of potential instability in Iran is the slowness with which the government has addressed the problem of economic reform. Unemployment is a severe problem, as is the decline in the productivity and export capability of the sectors of the economy that are not related to oil production. Inflation has been difficult to control.

The state subsidizes domestic energy consumption as well as food staples. It is also heavily involved in banking and other sectors of the economy. Meanwhile, the Bonyads,[75] nominally religious charitable organizations, are actually commercial enterprises that account for a large proportion of economic activity (some say as much as 50 percent). The Bonyads wield so much political power that they are beyond the range of government taxation and receive a variety of commercial privileges such as preferential foreign exchange rates.

President Khatami deliberately deferred economic reform until he had made progress on political freedom and the establishment of a rule of law in civil society. When he turned to Iran's economic problems, he sent the Iranian parliament proposals that would provide for substantial privatization and a partial flotation of the

---

74. "Iran's Economy, Ailing, Still," *The Economist*, August 14, 1999, 36.

75. These organizations are run by clerics and were initially set up to dispense the proceeds of the shah's nationalized assets to the poor. They have been accused of various forms of corruption but have thus far been protected from measures that would bring them under the control of government reformers. Their relationship with the center of power represented by the forces surrounding Supreme Leader Khamenei is especially close.

currency.[76] Creating employment opportunities is one of his most pressing tasks: close to 1 million jobs per year have to be created to accommodate a population that increased substantially during the 1980s and is now reaching adulthood.

Although there are Western-educated advocates of free market principles within President Khatami's inner circle, factionalism is rife. Khatami also must contend with supporters of state solutions to economic problems. It is not yet clear whether the Iranian effort to implement economic reforms will be successful and whether reforms will occur quickly enough.

The problem of the debt Iran has incurred to cover its budget deficits is severe and was exacerbated by the low oil prices of 1998 and 1999, but it should be alleviated somewhat by the 1999 price restoration. Resolution of Iran's debt problems has been further complicated by the U.S. policy of discouraging rescheduling by Iran's creditors.

Iran's economic problems will worsen over the next 20-year period if it is unable to make a successful transition from its current dependence on oil income.[77] Iran is strenuously trying to build up its capacity to produce natural gas. The ability of Iran to export natural gas will become crucial to Iran's economy. The urgency of completing the oil-to-gas transition is reinforced by the fact that Iran's baby boom is expected to reach the labor force at the same time as the country's oil exports and income begin to fall.[78] If this does in fact happen, unemployment and social discontent can be expected to grow at levels that could trigger serious political instability. Over time, the intersection of demographic pressure and a failed economy can lead to an explosive political situation.

The European View

Europe's assessment of Iran's internal stability is that, among the gulf countries, only Iran has demonstrated an ability to change its political leadership through a relatively democratic electoral procedure. The process of democratization is continuing to strengthen, albeit within the constraints of a constitution that divides power and provides elaborate checks and balances on the freedom of each branch of government.

Iran's efforts to establish conditions for foreign investment and rebuild its economic interdependence with OECD countries have so far been limited by the standoff in U.S.-Iran relations. If and when the standoff eases, Iran, with a more democratic system of government than any in the gulf, with a large population with a high skills potential, and with a relatively diverse economy, could become an

---

76. The International Monetary Fund (IMF) has recommended a full currency devaluation for Iran but this is politically dangerous because it will raise domestic prices and destroy the economic safety net that is a collateral effect of an overvalued currency. See "Quarterly Report—Iran," *Middle East Economic Digest,* October 26, 1999.

77. Although Iran often expresses concern about running out of oil and ceasing to be an oil exporter, it currently has 93 Bbbl of reserves.

78. "The Greater Middle East," *1998 Strategic Assessment: Engaging Power for Peace* (Washington, D.C.: National Defense University), 5. <http://www.ndu.edu/inss/sa98/sa98ch4.html> (accessed July 5, 2000).

important source of stability and development in the gulf. Unfortunately, this road is currently blocked.

Nigeria

With President Olusegun Obasanjo's inauguration in May 1999, Nigeria returned to elected civilian rule and moved forward in an unexpectedly promising—though still deeply problematic—turnaround of one of Africa's most powerful states and its most vital producer of both oil and gas. Obasanjo's ascendance followed an era of egregious misrule by a northern-dominated military regime that systematically plundered the state, including the central oil sector, pushing Nigeria into international pariah status and the nation-state itself to the edge of internal breakdown. The first six months of Obasanjo's tenure saw initial movement on human rights, reform of the military, management of the economy, and renewed dialogue with the World Bank and IMF. Since the end of 1999, however, progress has stalled as it has become apparent just how difficult it is to reform the military, end pervasive corruption, achieve timely debt relief, and ameliorate Nigeria's stark ethnoregional tensions.

Beginning in 1998, the Clinton administration made Nigeria one of four global priority transitions and followed up with successive secretary-level visits to Nigeria, a state visit by President Obasanjo to Washington, and a threefold expansion of U.S. assistance programs (more than $100 million in FY 2000). These efforts were in recognition of the exceptional nature of Nigeria's recovery from the Abacha era, its centrality to West Africa's stability and economic future, its special and rising stakes for U.S. energy security and effort to combat transnational crime, and its continued essential contribution to conflict resolution and peacekeeping in Sierra Leone and elsewhere.

At present, Nigeria accounts for 8 percent of U.S. oil imports and an estimated $7 billion in investment in the energy field. Major offshore oil finds in 1999 and plans to develop a regional gas pipeline to serve three neighboring West African countries (at an estimated $500 million) will further elevate U.S. energy stakes in Nigeria in the coming years.

Obasanjo will need to demonstrate substantive progress in three core areas to reduce the present high risk of future instability in Nigeria.

Reform of the military

Early in his tenure, President Obasanjo dismissed 200 senior officers and began deliberations over downsizing the military from its present level of 70,000–80,000 (high numbers are probable ghosts still on payroll) to a force of approximately 40,000. Another round of early retirements and dismissals is expected in mid-2000. In the meantime, Nigeria has had to accommodate the return home of upwards of 6,000 soldiers who had served for extended periods under harsh conditions in Sierra Leone.

The military feels it must address dilapidated housing, transport, and communications, but the government has had difficulty winning agreement from the national assembly on sufficient funds to implement an effective reform program. Senior military leaders still resist the uncertain and incomplete assertion of civilian control over the military, and the loyalty and influence of retired officers remain open questions. When conflict flared in the oil-rich Niger Delta in late 1999, the military was called back in yet again to assert, rather ruthlessly, local control. If ethnoregional violence should escalate sharply, pressure will mount—as in past moments of national crisis—to turn to the military to guarantee elementary law and order.

Management of ethnoregional demands

Escalating communal violence is the most conspicuous problem facing the Obasanjo administration. Riots that flared in Kaduna in February and March 2000 left 1,500 dead, according to official estimates. This was the gravest instance of concentrated interethnic violence in decades and shocked the nation. In 2000, disputes have intensified over shari'a law in northern states, the distribution of power between states and the central government, and national policing policy. Pressure has mounted for a sovereign or constitutional conference although that is a subject that divides the nation along ethnic and North-South lines. The Obasanjo administration has responded to the shari'a law challenge in northern states with considerable caution, and its overall policies toward deep enduring ethnoregional tensions remain vague.

Debt relief and improved governance

Under the best of circumstances, it will take a concerted effort and at least another one to two years to begin to reduce the sizable drag on the national economy posed by Nigeria's $30 billion debt. Efforts to date to reduce corrupt practices in the oil sector, the ministries, and the budgeting process have yielded minimal results.

Before Nigeria can go to the Paris Club to appeal for debt rescheduling, it must first reach agreement on an IMF-funded program and pass through a period of disciplined economic management. Negotiations with the IMF have been subject to constant delays, and considerable domestic political opposition remains to the IMF and its perceived encroachment on sovereign prerogatives. If these obstacles are overcome, Nigeria will have to demonstrate disciplined macroeconomic management for an indefinite period before it can turn to the Paris Club to appeal for debt relief.

Iraq

Iraq probably holds close to 11 percent of world oil reserves, which is second only to Saudi Arabia's 25 percent share. Because Iraq is also a very low-cost producer, the ability of Iraq to continue to supply oil to world energy markets is crucial. Internal instability can disrupt the current supplies and present an inhospitable environment for effective investment in the expanded production capabilities that will be required to keep pace with the expected surge in world oil demand during 2000–2020.

Saddam Hussein has been ruling Iraq with an iron hand, and it is difficult to imagine a scenario whereby externally based opposition groups would be able to link their efforts with forces within Iraq that are strong enough to overthrow him. Yet the probability that Saddam will pass from the scene within the time frame 2000–2020 is very high. When this happens—whether peacefully or violently—Saddam will leave behind a country that has been isolated by international sanctions and traumatized by severe economic deprivation as well as by the various brutalities of the current Iraqi regime. Moreover, the world has no assurance that Saddam's successor will not be his mirror image.

Three factors will make any transition to a new regime fraught with danger as far as the prospects for political stability are concerned.

- First, prominent ethnic and regional divisions separate the Kurdish North and the Shi'a South from the Sunni center that currently controls the country—through rule by the secular Baath Party that represents only a minority of the country. These divisions provide a tempting target for neighboring states that may seek to exploit Iraq's internal problems to weaken the country. The Shi'a revolt of 1991 came as a surprise to the West, even though the Shi'ites themselves believed that the West would react favorably. Similar surprises could be in store, and any serious internal conflict would likely have the same disruption potential for Iraq's oil exports as the Iranian revolution had in Iran.
- Second, there is a leadership vacuum caused in part by Saddam Hussein's deliberate policy of preventing any competing center of power from gaining enough strength to threaten his regime and also by inherent tensions among competing clans. Because sanctions and authoritarian rule have interfered with natural political and economic evolution and because Iraq exhibits the characteristics of a petro-state described on page 51, no institutions other than the military have developed in ways that could provide leadership alternatives and a unifying structure of political loyalties to unite the country and contribute to its effective governance. In 2000, the entire civil service of Iraq is associated with the Baath Party, and if the party is overthrown there will be an administrative vacuum as well.
- Third, Iraq will face a heavy reparations burden unless a formula of partial forgiveness is devised to support a new regime. Some observers calculate that Iraq would have to pay reparations of $2 billion every year for the next 100 years, a yearly rate equal to 20 percent of Iraq's 1990 preinvasion oil income, before the reparations bill would be paid. Many Iraqis already tend to blame the West for the economic deprivation caused by the international sanctions regime. A continued heavy burden of sanctions could contribute to a spirit of revanchism that could have destabilizing regional implications. Moreover, even though Iraq is a secular state, it is still possible that an Islamic fundamentalist movement could gain strength by capitalizing on continued economic deprivation and Iraqi attitudes of hostility toward the West.

The Iraqi regime has defined extensive opportunities for foreign investment in existing and undeveloped Iraqi oil fields under production sharing terms that are not matched in Saudi Arabia, Iran, or Kuwait. The rapid development of the oil sector will be the key to economic success for any regime in Iraq, and foreign oil companies will find Iraq attractive and desirable as soon as the international situation permits.

Mexico

Although there are many reasons to be optimistic about Mexico[79] at the turn of the twenty-first century and beyond, it is not difficult to construct a plausible scenario of social chaos that could lead to an alteration of the constitutional order and, perhaps, a disruption in oil supply from the world's fifth largest producer and tenth largest exporter.

There have been many signs of change and progress in Mexico in recent years. Primary among these have been an opening up of the political system and the conclusion of the North American Free Trade Agreement (NAFTA) between Mexico, the United States, and Canada. NAFTA has generated more than $174 billion annually in two-way trade between the United States and Mexico.

The revision and strengthening of the country's institutions are ongoing processes, however, and economic reforms have been only partially consolidated. The first test of Mexico's ability to sustain its ongoing democratic revolution was passed successfully in early July 2000, when the ruling Institutional Revolutionary Party (PRI) was ousted from power for the first time in 71 years. The vote against the PRI was clear-cut and not contested. Migration of the disaffected to the United States continues to act as a safety valve but with little impact on internal pressures.

Several other factors are potential contributors to Mexico's political instability.

- Large segments of Mexican society continue their opposition to the economic policies since 1989, and recent attempts to liberalize the economy have not yet benefited a majority of the Mexican people.[80] The gap is widening between the prosperous North and the impoverished South, which could form the basis for future unrest.
- Mexican politics has become polarized. The emergence of a revolutionary figure cannot be discounted if legislative gridlock develops and the political parties engage in endless bickering. The political party system is weakening, and the patience of the electorate is not endless. Nonetheless, the election of Vicente Fox as president has given renewed optimism to the Mexican political process. Now, can he deliver what he has promised?
- Mexicans expect continued political violence and harbor a feeling of personal, physical insecurity brought on by serious crime waves in many cities and an almost complete distrust of the police.

---

79. Lowell R. Fleischer, CSIS, contributed this analysis.

80. There is "crushing Mexican poverty (85 percent of Mexicans are either unemployed or not earning a living wage)." David, "Saving America from the Coming Civil Wars," 103–118.

- The severe problems of pervasive drug trafficking and the consequent influence of drug money in political and economic niches of Mexican society continue to dominate—and contaminate—relations between Mexico and the United States.

But beyond these concerns, failure to follow through on the badly needed transformation of the Mexican energy sector, especially electricity, could reduce Mexican future export flows. Without more private participation and competition, the oil sector—dominated by the Mexican national oil company PEMEX, that provides fully 35 percent of federal revenues—will not attract the huge investments it needs.

CHAPTER 6

Political Instability in Oil-Exporting Countries

Each top oil exporter—Iran, Iraq, Saudi Arabia, Kuwait, the UAE, Algeria, Libya, Venezuela, and Nigeria—is unique in its politics, society, ethnicity, and economics. Yet each country also falls within a subset of countries described as petro-states.[81] Some analysts believe that the high dependence on petroleum revenues in these states has caused them to experience a distorted form of political and economic development that, in turn, makes them especially vulnerable to the possibility of future internal strife.[82] This is especially true if the process of modern political and economic development in a specific country did not begin until the simultaneous discovery and exploitation of oil.[83]

Special Characteristics of Oil Production

Extracting petroleum for export, like the process of extracting other forms of mineral wealth, does not inherently lead to other forms of economic activity in the country in which the extraction takes place. The special characteristics of the petroleum sector prevent it from having linkages with other sectors of the economy. Oil is produced in enclaves and moved to domestic consumers via pipelines although small amounts move by rail and by water. Most exported oil is moved by sea.

Pipeline transport, although efficient, does not provide transportation infrastructure and the economic development that the construction of road and rail networks provides to otherwise undeveloped areas of a newly developing country. And oil production, unlike agricultural production or manufacturing, provides no multiplier effect for general economic activity in the country or even in the geographically contiguous local area. Moreover, the process of producing oil is technologically advanced and capital intensive, requiring a highly skilled—but rela-

---

81. Although it may be premature to characterize Azerbaijan, Kazakhstan, and Turkmenistan—all key players in the Caspian—as petro-states, they are already exhibiting characteristics that place them in this category.

82. Many ideas presented here are drawn from Terry Lynn Karl, *The Paradox of Plenty: Oil Booms and Petro-States* (Berkeley, Calif.: University of California Press, 1997), and Jahangir Amuzegar, *Managing the Oil Wealth: OPEC's Windfalls and Pitfalls* (London: I. B. Tauris Publishers, 1999).

83. If the discovery of oil occurred after a certain degree of economic diversification and political development had already taken place, oil exploitation would have played a less defining economic and political role.

tively small—workforce. Therefore, the petroleum sector cannot contribute much to national employment goals in countries with rapidly growing populations. Oil-producing states, therefore, have had to make special efforts to use their oil income to sow the seeds of economic diversification—usually with disappointing results.

State Control of Oil Wealth

In petro-states, the wealth that is derived from oil production is immediately transferred to the state, which then determines the allocation of the oil wealth to various groups and regions within the country. This is true even in oil-producing countries that have not chosen to follow a socialist path of economic development, but it has resulted in the same economic inefficiencies and distortions that have occurred in socialist command economies. Because of their control over oil revenues, governments that are inefficient, corrupt, and wasteful are able to remain in power long after they would have been seriously threatened in more normal political circumstances.[84] More important, perhaps, is the fact that the private sectors of these countries become dependent on state guidance, credit, protection, and subsidies, and they tend to lose—or fail to develop—the dynamic, innovative, risk-taking, and independent spirit required for a healthy nation-building effort.[85]

Lack of Political Accountability

Instead of deriving its revenue primarily from taxation of its own citizens[86] and from taxation of a variety of domestic economic endeavors, the petro-state essentially derives its wealth from the international community, through its exports abroad, and distributes the wealth to its domestic constituents.

The distribution of wealth has an ad hoc quality to it. Wealth is transferred to individuals, groups, and regions—in the form of patronage, subsidies, and credits. The petro-state government provides not only generous public works programs and social safety-net benefits, including artificially low prices charged for subsidized domestic oil and basic foodstuffs, but also wages for unnecessary employment, funds for bloated bureaucracies, and support for graft and corruption.

Petro-state governments do not have to extract their budget revenues from domestic taxpayers. The various forms of ad hoc state payments are intended to buy political loyalty and social peace, and they may temporarily have that effect. But without a system of domestic taxation, there is no parallel system for holding the government accountable for the way in which the oil wealth is spent.

In countries without established democratic institutions, the government's access to oil or gas export revenues enables it to avoid or postpone the development

84. Amuzegar, *Managing the Oil Wealth*, 207.

85. Ibid.

86. Some petro-states have no personal income taxes at all; in others, tax avoidance is tolerated at dimensions that render nominal tax rates relatively meaningless.

of democratic institutions that, if not abused, might be expected to provide a means of adjusting policies and accommodating and reconciling competing domestic interests. The lack of public accountability helps to lay the groundwork for future political instability; the country's people have no institutions to guarantee that their voices are heard.

High Levels of Debt

The contrast is striking between the level of OPEC-government revenues in 1974—when OPEC revenues increased 11-fold compared with the years immediately preceding—and the level of government debt just 20 years later. By 1994, Algeria had an external debt level of $30 billion; Indonesia, $108 billion; Iran, $23 billion; Nigeria, $33 billion; and Venezuela, $37 billion.[87] The total debt service bill that year for these five countries exceeded $28 billion.[88] Although there have been some recent changes on a country-by-country basis, the updated statistics are similar. In 1998 Algeria owed $31 billion; Indonesia, $151 billion, Iran, $14 billion, Nigeria, $30 billion, and Venezuela, $32 billion; and the total debt service cost was $34 billion.[89]

The financial and attendant political dilemmas of the petro-states are compounded by the fact that their debt obligations and the spending requirements necessary to forestall internal turmoil deny them the option of significantly lowering their prices to maintain market share. They cannot follow the rules of classic cartel behavior; they cannot lower their prices to prevent new entrants from coming into the market, only to raise them again after potential new competitors are forced away. Thus, in the long run, the total level of income that petro-states can gain from their oil is much lower than it would be otherwise.

Boom-and-Bust Economic Cycles

An important source of political instability derives from the boom-and-bust cycle that characterizes the international petroleum market. During periods of windfall profits, the leaders of petro-states initiate highly ambitious government projects. Several factors motivate them: the sudden availability of large sums of money; the desire to satisfy national pride; and, in oil-producing countries that lack high levels of oil reserves, the feeling that quick action is necessary to reach the goals of economic development and diversification before the oil income runs out. The scale of the projects initiated during boom periods requires continued high investment expenditure over long periods of time.

During every boom, groups that have been receiving state benefits vie with each other to gain a larger share of the new supply of state resources. In fact, some observers note that dramatic instances of political instability in petro-states tend to

---

87. *World Development Indicators 2000* (CD-ROM) (Washington, D.C.: The World Bank, Development Data Group, 2000). The most recent data reported are for 1998.

88. Ibid.

89. Ibid.

occur following sharp increases in world oil prices.[90] When prices plunge during the bust period, the petro-states need additional income to fill their revenue gap, in part because of the long-term financial obligations they have incurred. Additional income is usually obtained through international borrowing, often short-term and at high interest. Over time, the petro-states find themselves increasingly burdened by debt because they do not want to incur the political and social costs of implementing austerity measures during times of low oil income. When some (partial) efforts have been made toward economic reform, they tend to be abandoned as soon as oil prices rise again.

During low-oil-price periods, the petro-states seek to continue their social welfare and patronage expenditures at previous levels in the interest of maintaining social and political peace.

Thus, the cycle is one in which the petro-states get deeper and deeper in debt and run greater and greater risks of internal upheaval when the expectations of their populations can no longer be met. Islamic fundamentalism appeals to people who are economically deprived. Population growth in the petro-states has shot upward, and the number of restive unemployed young people has increased substantially.

A paradox exists. For oil-importing countries, the risks of disruption in one or more exporting areas are best mitigated by the existence—and use—of spare oil producing and exporting capacity elsewhere. The existence of spare capacity destabilizes the market, however; it tempts exporters with boom-and-bust commodity competition. Spare oil producing and exporting capacities are expensive to maintain. Exporters see little reason to maintain spare capacities if their goal today is to maximize price—as it is—and not market share—as it was. National interests may in the future dictate that market share is more important than price. At that time it would be practical for exporters to invest in developing spare producing and exporting capacities, first to protect and then to expand market shares.

The possibility does exist of developing an investment cycle for oil that is typical of other commodities: alternating inadequate and surplus supplies. Does this matter? The major importing countries' economies are large and diverse enough to cope with cyclic fluctuations in oil import prices. The major exporting countries' economies are not. Therefore, investment cycles do threaten the political stability of exporting countries. This is the basis of the long-standing call by exporting countries for some guarantee of markets and some stability in international oil pricing. But current economic ideology in the main importing countries argues against any attempts to intervene in commodity markets—even oil markets—to stabilize prices.

Geoeconomics and Internal Instability

Major oil exporters and consumers are currently in a situation of extreme mutual dependence. The ability of a major and prolonged supply disruption to trigger a

---

90. Ibid., 193–196.

world recession was described on page 18. But a reverse relation is also worthy of attention. The major oil exporters are each so heavily dependent on oil revenue that any substantial reduction in oil revenue could hasten the very political upheavals that can cause disruption of oil supply to the rest of the world.[91] It is a worst-case circular relationship whereby

- a severe and prolonged oil supply disruption caused by factors other than the world oil market triggers a world economic recession that decreases consumer demand for oil,
- the reduction in consumer demand causes such a drop in oil revenue that the top oil exporters would become susceptible to internal political instability that could cause a further disruption of supply.

A recession of worldwide proportions that causes oil consumption to decline everywhere would cause financial dislocation in all 40 or so of the countries that rely on export income derived from oil. Thus the possibility that a world economic recession could occur constitutes a geoeconomic risk that has both energy and geopolitical implications. And, because recessions usually last much longer than oil disruptions, it could be argued that the threat posed by a macroeconomic downturn is actually more serious than an oil disruption.

Angola: A Future Concern

Angola is perhaps the perfect example of a nation plunged into despair because of the tragic manipulation of its natural resource endowment.[92] Angola has chosen to employ the income derived from exports, particularly of oil and diamonds, to fund an ongoing civil war. Oil exports support the Popular Movement for the Liberation of Angola (MPLA), which has controlled the government since Portugal granted independence to Angola in 1975. Diamonds fund the National Union for the Total Independence of Angola (UNITA), which stands in opposition. Except for four years, between 1994 and 1998 when the United Nations oversaw a peace accord, these two groups have been engaged in a struggle that has devastated the country. Perhaps as many as 500,000 have been killed during the civil war and the refugee count reaches almost 2 million.

Angola today is literally in ruins. The capital city of Luanda is chronically short of water and electricity. Indeed, it is estimated that only 15 percent of Angola's population have access to electric power. The life expectancy at birth for the population as a whole barely exceeds 48 years, and it is judged that only 4 out of 10 Angolans age 15 and older can read and write. Subsistence agriculture provides the main live-

---

91. Michael C. Lynch makes this point in "Oil Scarcity, Energy Security and Long-term Oil Prices—Lessons Learned (and Unlearned)," *International Association of Energy Economists (IAEE) Newsletter* (Third Quarter, 1999).

92. Jon Lee Anderson, writing in the *New Yorker,* provides a sobering insight into Angola's deterioration. His article carries a very apt subtitle: "One of the biggest humanitarian disasters in Africa has been created by a war paid for with oil and diamonds." See Jon Lee Anderson, "Oil and Blood," *New Yorker,* August 14, 2000, 46.

lihood for 85 percent of the population.[93] Much of the food must be imported because millions of land mines infest tillable areas.

Oil has been produced in Angola for almost 50 years, but it was not until the 1960s that the industry came into its own with a discovery offshore the enclave of Cabinda. Growth in production has been steady but not spectacular over the years, with output reaching 475,000 bbl/d in 1990 and 780,000 bbl/d in 1999. Angola is not a member of OPEC, nor is it presently a ranking non-OPEC producer and exporter. The United States is the most important buyer of Angolan oil, with these purchases reaching a peak of 465,000 bbl/d in 1998.

It is not Angola's oil past that is of interest, however. Rather, it is the country's oil future that commands attention, a future made possible by recent advances in deepwater exploration and production techniques. The Angolan government expects that oil production will increase to 1 MMbbl/d by the end of 2001 and to 1.4 MMbbl/d by 2003.[94] Other estimates underscore the future importance of Angolan oil in noting that for the five-year period, 2001–2005, Angola will provide slightly more than one out of every four barrels in crude oil production growth among non-OPEC countries.[95] The oil is of good quality, production costs are low, and the U.S. East Coast offers an attractive and easily accessible market.

To date, the civil war has not disrupted oil exploration and production activities, in that the major fields are offshore Cabinda and isolated from the fighting, but downstream (i.e., refining and marketing) is in almost total disrepair. The country's one refinery is old, small in capacity (just 39,000 bbl/d), and badly in need of renovation. Similarly, most of the natural gas extracted in the country is burned off for lack of a market.

If the past is prologue, then little of the coming oil wealth will pass to the benefit of the Angolan people. Rather, the hard currency earned from oil exports more likely will be employed in support of efforts by the MPLA to subdue all opposition, and that in turn can only translate into further devastation. Then, if the projections of future expansion in oil production are reasonable and as Angola supplies ever-increasing volumes to the world oil market, the domestic situation takes on international implications. All the unwelcome attributes of a petro-state can be found in Angola.

---

93. *CIA World Factbook 1999*, "Angola" (Washington, D.C.: Central Intelligence Agency) <www.odci.gov/cia/publications/factbook/ao.html> (accessed August 9, 2000).

94. *Angola (Country Analysis Briefs)* (Washington, D.C.: Energy Information Administration, U.S. Department of Energy, July 2000) <www.eia.doe.gov/emeu/cabs/angola.html> (accessed August 8, 2000).

95. Deutsche Banc Alex. Brown estimates, June 27, 2000.

CHAPTER 7

Cooperation for Energy Preparedness

Canada

The neighbor of the United States to the north, Canada, has been quietly but gradually assuming a growing role as a major supplier to the United States of both oil and natural gas. No other exporter can claim that distinction. At the beginning of the 1990s, the United States imported an average of 8.02 MMbbl/d of crude oil and petroleum products. Of that amount, 934,000 bbl/d or 11.9 percent originated in Canada. As the decade came to a close, in 1999, Canadian oil deliveries to the United States averaged slightly more than 1.5 MMbbl/d, accounting for 14.3 percent of total oil imports.[96]

Canada also sells natural gas to the United States, supplying in 1999 approximately 95.6 Bcm and accounting for virtually all natural gas imports delivered by pipeline (a very limited volume comes from Mexico).[97] Small volumes of LNG, provided chiefly by Algeria, brought the total volume of natural gas imported that year to approximately 101 Bcm, meaning that the U.S. dependency on Canada for imported natural gas approaches 95 percent, a vulnerability ignored because of the shared mutual interests of Canada and the United States. Equally important, Canada has been taking an increasing share of the U.S. natural gas market, providing only 3.7 percent of total gas consumed in 1980, 7.7 percent in 1990, and just under 16 percent in 1999.

Thus, if imports of oil and natural gas are totaled, Canada becomes the single largest provider of energy to the United States, exceeding Saudi Arabia, Venezuela, or Mexico.[98]

Canada is poised to expand sharply its exports of oil to the United States in the coming years. If all the announced oil sands projects are brought to fruition, by 2010 Canadian crude oil production will be approaching 3.5 MMbbl/d, compared with approximately 2.1 MMbbl/d in 1999.[99] Another estimate, by Canada's National Energy Board based on current trends, envisages oil production reaching 2.6

---

96. EIA, *Monthly Energy Review* (March 2000): 52.

97. Ibid., 75.

98. James Burke, "Oil Industry Rebounds in Canada as It Becomes Top U.S. Supplier," *New York Times,* June 12, 2000, p. 1.

99. Dan Whelan, director general, Energy Resources Branch, Natural Resources Canada (remarks at the 27th international energy conference on identifying energy-sector drivers in the coming decade, Boulder, Colo., April 17, 2000).

MMbbl/d in 2010. Future growth under either scenario will be based on the oil sands of western Canada, where reserves are currently placed at 305 Bbbl. The bulk of any increment in Canada's production will be directed south, to buyers in the United States. A number of concerns and challenges—economic, environmental, and social—must be overcome, however, if oil sands development is to proceed at the desired pace.

Oil production in Canada is centered in the West, consumption centers are in the East, and no pipeline directly links the two. Therefore, oil consumed in Canada's East is imported. What are the implications if for some reason these imports are disrupted? Would exports to the United States be redirected to Canadian consumers? No, because NAFTA requires that any shortages would be shared between the two countries.

There can be no more secure supplier to the United States than Canada. U.S. consumers should welcome any expansion in Canadian oil, as each additional Canadian barrel imported will in its own way reduce the perceived U.S. vulnerability.

Energy Emergency Preparedness

The economic well-being of the United States and its allies and trading partners remains vulnerable to the damaging effects of an oil supply interruption. The experiences of the 1973–1974 oil embargo, the effects of the Iranian revolution in 1978–1979, and the Iraqi invasion of Kuwait and its aftermath in 1990–1991 provide clear evidence of the link between energy and national security.

A cornerstone of the IEA emergency response mechanism is the requirement for its members to maintain an emergency oil stockpile program under government control. The stockpiling requirement calls for each of the 25 IEA members to hold oil stocks that are equivalent to at least 90 days of net oil imports. These government-controlled stocks are essential for the successful activation of the emergency oil sharing systems developed by the agency in the 1970s and 1980s. The IEA emergency response mechanisms also include demand restraint, fuel switching, and surged oil production.

In the United States, the strategic petroleum reserve (SPR) was established by legislation in 1975, and purchases began in 1977. Although the original legislation called for establishment of a 1 Bbbl, the government-owned crude oil stockpile in 1999 contained only 575 MMbbl—the equivalent of close to 60 days[100] of U.S. net oil imports. The policy of the United States regarding severe oil supply disruptions is to rely on market forces to allocate supply and to supplement supply by the early drawdown of the SPR in large volumes and in coordination with our allies and trading partners.

Outside of the IEA very few countries have established energy emergency response mechanisms. The EU has an oil stock requirement but all of its members

100. U.S. net oil imports are calculated to be 9.8 MMbbl/d on the basis of data presented in *BP Statistical Review of World Energy* (June 1999): 7, 10.

are also members of the IEA. APEC is actively considering establishing energy emergency response mechanisms, including the possibility of initiating an APEC oil stock program. This is especially significant because 16 of the 21 APEC member economies are not members of the IEA; yet these economies are projected to account for a significant share of the rapid growth in Asia's net oil imports during 2000–2020. Another approach would be for the IEA to expand participation in its stock-sharing mechanism to encompass interested nonmember countries.[101]

101. The developing countries, however, may resist being enfolded into an industrialized-world institution such as the IEA. One alternative would be the development of an institution for stock sharing and emergency preparedness that is open to all countries on an international basis, but many barriers would have to be overcome before such a solution could be implemented.

CHAPTER 8

Geopolitics, Energy, and the Environment

During 2000–2020, the geopolitical aspects of the energy use–environment relationship are expected to become more apparent.[102] Continued economic growth worldwide will be at the heart of the issues at stake, especially in the developing world. Differences in interest between countries that are already industrialized and countries that are either beginning or continuing the process of industrialization will intensify. More people will realize that international cooperation will be necessary to resolve the key environmental issue of climate change.

In many parts of the world, the issue of the impact of energy production and consumption on global and regional environments has begun to displace issues of energy security and price stability.[103] In the United States in 2000 little or no attention is paid to the issue of security of supply, but global-warming concerns attract heavy media coverage. Note also that

- Energy consumption is intertwined with the process of economic growth;
- Fossil fuels provide more than 86 percent of the energy that is currently consumed; and
- Many environmental problems are associated with the production, use, and transportation of fossil fuels.

Projections for 2000–2020 indicate that energy use will increase by more than 50 percent. Unless there is a dramatic technological breakthrough that would provide economically attractive alternatives to fossil fuels—and government-mandated[104] use of the alternatives—fossil fuels will continue to provide the greatest part of the world's energy.[105]

Projections also indicate that the 83 percent of the world's population that live in nonindustrialized countries will double their current energy consumption. These countries will be engaged in their own processes of industrialization during the coming 20 years—at the same time that their populations will be growing at a marked rate.

---

102. See the companion SEI report, volume 2, *The Supply-Demand Outlook, 2000–2020,* for a detailed discussion of the issues involved in the relation between energy use and the environment.

103. As indicated, for example, in the EIA's *International Energy Outlook 1999.*

104. Government mandates would be at odds with the concept of reliance on market mechanisms for the solutions to economic problems and are therefore unlikely prevail.

105. Fossil-fuel use compared with use of other sources of energy will increase to 88 percent by the year 2020.

For an assessment of the geopolitical impact of environmental issues, they can be divided into two categories: air and water pollution that is close at hand and pollution in the upper atmosphere. Industrialized and developing countries have different perspectives on these environmental issues.

- Environmental problems at ground level and in the lower atmosphere
  - Industrialized countries have begun to take steps through a combination of regulation and technological innovation to address soil and water pollution and pollution of the lower atmosphere (urban ozone, acid rain).
  - Developing countries are becoming increasingly aware of lower-atmosphere degradation as their citizens experience intense levels of air pollution in megacities and acid rain that crosses national boundaries.

    While they struggle to increase their standards of living, these countries find that their GNP is reduced annually by an estimated 2–4 percent because of lost workforce productivity from pollution-caused illnesses.[106]

    But the developing countries lack the technological and economic resources—clean-coal technology, for example—to mitigate lower-atmosphere pollution and often assert that technologically advanced countries should provide them with these resources on a grant-aid basis.
- Upper atmosphere pollution and climate change
  - Industrialized countries are undertaking various (partial) national measures to reduce the levels at which carbon dioxide and other greenhouse gases are released into the atmosphere. They have also undertaken binding commitments under the terms of the Kyoto Protocol[107] to reduce their carbon emissions to levels that are several percentage points below those that existed in the benchmark year 1990.[108]
  - The developing countries have thus far refused to undertake Kyoto Protocol commitments of their own although they are participating signatories. They believe that the reductions in fossil-fuel use that would be necessary to achieve the Kyoto Protocol carbon-emissions-reduction targets would substantially constrain their efforts to continue along the path of economic development. They argue that the industrialized countries' previously accomplished industrialization has caused the climate change problem and, therefore, industrialized countries should demonstrate a good-faith effort

---

106. For more information, see <www.wri.org>.

107. The events leading up to the 1992 UN FCCC (the Rio Treaty) and the subsequent protocol signed in Kyoto are described in detail in the SEI companion document, volume 2, *The Supply–Demand Outlook, 2000–2020*. The debate about the merits of the protocol is summarized and evaluated as well.

108. As Annex I countries, the industrialized countries took on these commitments by signing the Kyoto Protocol in November 1997. None of the Annex I countries, however, has yet ratified the protocol although the EU countries are expected to do so and the countries of the FSU might follow suit. Some observers believe that the EU will not ratify in the absence of ratification by the United States.

to reduce their own emissions[109] before asking the developing countries to do so.

The United States tried to achieve developing-country emissions-reduction commitments during the Kyoto Protocol negotiating process but was unsuccessful. Meanwhile, the U.S. Senate is on record that it will not ratify the Kyoto Protocol unless it is changed to include developing countries in the emissions-reduction process. President Clinton stated that he would not submit the protocol for ratification until this goal is met.

It is not clear whether—or how—the impasse between the U.S. position and the developing country position will be resolved. In addition, many implementation details of the Kyoto Protocol still must be negotiated although some progress is being made. Therefore, the Kyoto approach to addressing the problem of climate change cannot be counted on to have a significant impact on climate change during 2000–2020.[110]

Some believe the Kyoto approach is flawed because it represents a politically attractive quick fix to the problem of climate change. It allows leaders of industrialized countries to burnish their environmental credentials by lending Kyoto their support. But the measures mandated by the Kyoto Protocol—even if implemented—would not be effective enough in reducing long-term levels of carbon emissions to have a meaningful result.[111]

Instead, it is argued that scientifically advanced countries should undertake significant R&D efforts with the goal of achieving a technological breakthrough that would either permit continued fossil-fuel consumption and economic growth without damage to the environment,[112] or provide cost-competitive energy alternatives to fossil fuels. This approach is based on two assumptions:

- The climate change problem is very serious, and
- Its ultimate solution requires a long-term focus.

The future geopolitical impact of the issue of climate change is enhanced by the fact that it will have different effects on different countries,[113] which could become apparent during 2000–2020. Although the specific effects of climate change cannot be predicted with accuracy because complex relationships are involved,[114] it is expected that the northern temperate zone of the world will experience a net bene-

---

109. The United States, for example, accounted for more than one-third of worldwide carbon emissions in 1997 when the Kyoto Protocol was signed.

110. The problem will be partly ameliorated as Annex I countries begin to establish regulations to help them meet their Kyoto Protocol emissions-reduction targets.

111. Carbon emissions have a very long chemical life. Consequently the beneficial effects of current reductions will not be evident for two to three generations.

112. Methods of carbon sequestration, for example, would permit combustion of fossil fuels but would trap and dispose of the collateral carbon.

113. And it will affect different regions within large countries that encompass different climate zones—the Northeast and Southwest of the United States, for example.

114. For example, at the same time that temperatures increase, there may also be mitigating effects as a result of increased cloud cover and heat-reducing storms.

fit from global warming, while most of the developing world will find their already hot climates worsening.[115]

If this proves to be the case during the next 20 years, it will be an ironic paradox:

- The countries that will be receiving benefits from warming are the same countries that have historically contributed the highest levels of carbon emissions.
- Conversely, the developing countries that have thus far been unwilling to make emission-reduction commitments will suffer the most from global emission increases.[116]

The potential for climate change to increase the divisiveness of international political relations is clear, especially as countries become more aware of what climate change means to them in particular.[117] One danger from a global standpoint is that the developed countries will lose interest in climate-change negotiations when they begin to realize they will reap economic benefits from a warmer climate.[118]

Significant equity issues will be raised if global climate change imposes greater economic burdens on developing countries because of their geographical location. Such burdens would add a new layer of economic difficulty to those already in the developing world. The lack of development will make it more difficult for these countries to deploy climate-abatement measures that are available to the less-threatened developed world.

If countries in the developing world begin to experience effects of climate change that serve to increase their already difficult situation, the potential is great for revival of international political conflict that pits the industrialized North against the developing South. This happened during the early 1980s when the developing countries joined together to demand a new international order that would give them more favorable economic terms. The North–South controversy that is developing over the environment is likely to be divisive to an even greater extent and will make it much more difficult to resolve international problems that require cooperation between developed and developing countries.

---

115. One of the most important expected economic changes is in the field of agriculture: A warmer, wetter world with increased levels of atmospheric carbon dioxide is expected to be a "greener" world because carbon dioxide facilitates the process of photosynthesis. But in countries that become very hot, agricultural damage will take place. Thus it is predicted that agricultural production in the United States and other developed countries will increase, while in the developing world—where economies are much more heavily dependent on agriculture—levels of agricultural production will decrease. Robert Mendelsohn, *The Greening of Global Warming* (Washington, D.C.: AEI Press, 1999), 25.

116. Emissions include the prior contributions of the industrialized countries as well as the new emissions contributed by the developing countries themselves during their own process of industrialization during 2000–2020. See Mendelsohn, *Greening of Global Warming,* 3.

117. Ibid., 26.

118. Ibid., 30.

CHAPTER 9

Summary and Conclusions

The interaction between energy and geopolitics at the turn of the twenty-first century is likely to pose significant challenges to governments and policymakers around the world. Some of these challenges will resemble the challenges of past decades, especially insofar as geopolitical events are likely to affect the supply of energy.

The end of the Cold War, for example, has not removed the traditional geopolitical instability in the Persian Gulf region, nor has it led to a resolution of territorial disputes and regional rivalries in Asia. Conflict in either region could have important repercussions for the supply of energy, especially oil.

In addition, the proliferation of missile technology, WMD, and sophisticated and increasingly lethal conventional military systems means that any conflict could be far more destructive than in the past, causing far more severe energy disruption.

Oil-exporting states, with only a few exceptions, also face a complex mix of domestic economic and social challenges that energy-export revenues by themselves are incapable of resolving. Many of these petro-states remain vulnerable to internal political instability or cross-border subversion that could disrupt their ability to serve as reliable sources of energy, even as the world becomes increasingly dependent upon them to meet a constantly growing trend of energy consumption.

As important as geopolitical events are to energy, the world will also face significant challenges during the early decades of the twenty-first century from the impact of energy issues on geopolitical relations. In 1998–1999 the world witnessed how falling global demand for energy as a result of Asian economic turmoil caused major problems for key world energy producers such as Russia, Saudi Arabia, and Iran. Policymakers must remain attentive to the spillover effects on energy-producing countries of future drops in energy demand, especially when economic distress could quickly evolve into geopolitical upheaval.

Energy might affect geopolitics in 2000–2020 in two other principal ways. The first relates to the vast increases in energy imports that China and other Asian countries will require to fuel their economic growth. A large proportion of these imports will consist of oil and will have to be drawn from the gulf region, further increasing the geopolitical importance of that region as well as of the sea lanes over which energy supplies will be transported. As a result, the U.S. military presence in both the gulf region and Asia will increase in importance, creating opportunities for cooperation—and for tension—with both exporters and importers.

But Asia will also be increasing its consumption of natural gas, which will require the prior mobilization of large amounts of capital for the construction of a natural gas transportation infrastructure. New energy relationships are likely to be formed as a result. There is a natural complementarity between the energy resources in Russian Siberia and the Russian Far East and the burgeoning demand for natural gas in China and Southeast Asia. In addition, the countries of South and East Asia will be turning to the countries of Central Asia as potential new suppliers of energy.

The choices that the Asian countries make in terms of their future energy suppliers will have important geopolitical ramifications. Investors will not finance natural gas infrastructure, for example, without an expectation of reasonably long-term and stable political relations among the participating countries. But once the infrastructure is in place and the energy supplies begin to flow, an energy relationship will solidify the parallel political relationship. Thus it will matter geopolitically whether China turns to Russia or Central Asia as its primary regional energy supplier. Dependence on either (or both) will lessen Chinese dependence on the gulf.

Meanwhile, the need to confront global environmental challenges caused by energy use—climate change, for example—will lead to growing tensions between an already industrialized North and a rapidly industrializing South that cannot afford to place environmental policy ahead of economic growth.

Without doubt, the interplay between geopolitics and energy could have both dramatic short-term as well as protracted consequences for global stability and prosperity. Tensions between geopolitics and energy should come as no surprise to policymakers who have frequently been forced to deal with the consequences of this interaction. But appreciation of the underlying forces that connect geopolitical and energy developments and their likely future evolution will enhance the ability of policymakers to make timely and prudent decisions.

APPENDIX A

Energy Transit Choke Points

The United States and other countries have long been conscious of the risks posed by energy choke points.

- The Strait of Hormuz alone, for example, is a choke point affecting approximately 30 percent of world oil exports in the year 2000; this is expected to rise to 38 percent in 2020. On an average day, 14 MMbbl of oil move through the strait and there is no readily available alternative export route.

  Other choke points, if disrupted, would force temporary diversions to circuitous sea lanes, causing shortages of shipping and dislocation of supplies and prices unless spare capacity was available from other sources and put to work.

- Bab el Mandeb, a strait located between Djibouti and Eritrea (on the western shore of the Red Sea) and Yemen (on the eastern shore) sees oil in the amount of 3.3 MMbbl/d move from the Red Sea to the Gulf of Aden and on to the Arabian Sea and the Indian Ocean.

- The Bosporus, in Turkey, connects the Black Sea with the Sea of Marmara.[119] It is only one-half mile wide at its narrowest point and is one of the world's most difficult-to-navigate waterways. In 2000, 1.2 MMbbl/d per day move through the Bosporus, including oil from Russia and the Caspian region.

- The Suez Canal and Sumed Pipeline, in Egypt, connect the Red Sea and Gulf of Suez with the Mediterranean Sea. Crude oil moving northward, primarily from Saudi Arabia and Iran, goes through this route at a level of 3.1 MMbbl/d.

- The Strait of Malacca, between Indonesia and Malaysia, connects the Indian Ocean with the South China Sea and the Pacific Ocean. Oil destined for Japan, South Korea, China, and other Pacific Rim countries travels through the Strait of Malacca at a level of 8.2 MMbbl/d.

- The Panama Canal connects the Pacific Ocean with the Caribbean Sea and the Atlantic Ocean. Traffic traveling through the canal includes tankers carrying oil in the amount of 0.6 MMbbl/d.[120]

- Baku, in Azerbaijan, could eventually become a significant choke point if oil and gas from Kazakhstan and Turkmenistan as well as offshore in the Azeri

---

119. After transit through the Sea of Marmara, tanker traffic can proceed through the Adriatic Sea and on to the Mediterranean Sea.

120. *World Oil Transit Chokepoints* (Washington, D.C.: Energy Information Administration, U.S. Department of Energy, May 1998).

section of the Caspian Sea are moved by pipeline across the Caspian Sea and then sent through Baku en route to Western markets.

A P P E N D I X B

U.S. Sanctions That Affect Energy

Under the terms of U.S. legislation,[121] when a country is designated by the secretary of state as one that is engaging in a pattern of support for international terrorism, it is prohibited from receiving U.S. foreign assistance. The prohibition covers development assistance, economic support funds, agricultural assistance, participation in Peace Corps programs, and the provision of U.S. Export-Import Bank credits. The countries on the terrorism list in 2000 are Cuba, Iran, Iraq, Libya, North Korea, Sudan, and Syria. Third countries that are identified as giving assistance to countries on the terrorism list are also subject to the same sanctions.[122]

Libya

Separate U.S. economic sanctions against Libya, including an oil embargo, have been in place since March 1982 and were strengthened in 1986.[123] These sanctions against Libya are separate from the UN sanctions put in place as a result of the downing of a passenger airplane over Lockerbie, Scotland.

Iraq

U.S. economic sanctions against Iraq were imposed in August 1990,[124] immediately after Iraq's invasion of Kuwait. A complete trade embargo was imposed and financial assets of the government of Iraq that were located in the United States were blocked. These sanctions are separate from the multilateral sanctions imposed on Iraq by the United Nations.[125] Further sanctions were imposed in 1992 under the terms of the Iran-Iraq Non-Proliferation Act.

---

121. *Foreign Assistance Act of 1961,* as amended, Section 620(a).

122. *Foreign Assistance Act of 1961,* as amended, Section 620(g).

123. The 1982 sanctions against Libya prohibited imports into the United States of Libyan oil and placed U.S. exports to Libya under license; in addition, export licenses were denied for all goods on the national security list (that was applied, for example, to the Soviet Union). On January 7, 1986, in Executive Order 12543, these sanctions were broadened to include a total trade, investment, and travel ban while they extended the previous oil embargo to include prohibition against the purchase of oil (or other Libyan exports) by U.S. companies for export to third countries.

124. Executive Order 12722, August 2, 1990.

125. Executive Order 12724, August 9, 1990.

Iran

U.S. economic sanctions against Iran currently prohibit all trade and investment relationships although early in 2000 a general exclusion was granted permitting Iranian purchases of U.S. food and medicine and U.S. purchases of a very limited number of Iranian consumer items.

- An import ban, including direct imports of Iranian oil, was imposed by the Reagan administration in October 1987.[126]
- In 1992, Iran was made subject to the sanctions imposed by the Iran-Iraq Non-Proliferation Act.
- Sanctions against Iran were broadened substantially in May 1995 after the president determined that Iran had engaged in activities that represent "a threat to the peace and security of all nations [by its]...continuing support for international terrorism, including support for acts that undermine the Middle East peace process, as well as its intensified efforts to acquire weapons of mass destruction."[127]

  The timing of the sanctions blocked a major energy contract between Iran and a U.S. company, Conoco—an arrangement that was later taken over by Total, a French company.
- Sanctions constituting a secondary boycott were imposed one year later when the Iran and Libya Sanctions Act of 1996 (ILSA) was passed. This legislation prohibits companies that invest in the Iranian or Libyan oil and gas sector in amounts larger than $20 million from also conducting business in the United States.

Sudan

U.S. sanctions were imposed on Sudan in November 1997[128] because of Sudanese support of international terrorism and violations of human rights. Chevron, the major U.S. investor in Sudan, had already left the country because of a civil war there. Although not a major energy exporter, Sudan's oil resources are in the process of development, and the longest pipeline in Africa is being constructed on its

---

126. Ernest H. Preeg, *Feeling Good or Doing Good with Sanctions: Unilateral Economic Sanctions and the U.S. National Interest* (Washington, D.C.: CSIS, 1999), 47–71. Patrick Clawson, "Iran," in *Economic Sanctions and American Diplomacy*, ed. Richard N. Haass (New York: Council on Foreign Relations, 1998), 85, points out that the Reagan administration's decision to impose an import ban on Iran followed closely upon a unanimous vote in the Senate and a 407-5 vote in the House of Representatives criticizing the U.S. Department of Energy for having purchased Iranian oil for the U.S. Strategic Petroleum Reserve.

127. Letter from President Clinton to the Speaker of the House of Representatives and the President of the Senate, May 8, 1995, in transmittal of Executive Order No. 12959, effective May 7, 1995.

128. Executive Order 13067, November 3, 1997.

territory. Several companies from other countries—including CNPC, the national oil company of China—are currently operating there.

Afghanistan (Taliban)

U.S. sanctions were imposed on the ruling Taliban government of Afghanistan on July 4, 1999,[129] because of the protection that the Afghans allegedly accord to Saudi-born militant Usama bin Laden, who is accused of planning the August 1998 bombings of U.S. embassies in Kenya and Tanzania and suspected of planning subsequent terrorist attacks against U.S. facilities abroad.

Afghanistan is a key transit country for a proposed $1.9 billion, 900-mile gas pipeline that would carry gas from Turkmenistan to Pakistan. A U.S. company, Unocal, withdrew from the project in November 1998; both Pakistan and Turkmenistan are continuing their attempts to bring it to fruition,[130] but prospects are not favorable over the near term.

Nigeria

Although various proposals have been put forth for a ban on Nigerian oil imports and on new U.S. investment in Nigeria, they have never been implemented.

A variety of U.S. arms sales and economic sanctions have been imposed on Nigeria, dating back to the military takeover of Nigeria's civilian regime in 1993 but also concerning issues of airport safety[131] and the Nigerian narcotics trade.[132]

If, however, there is a reversion to military rule in Nigeria following the optimistic expectations engendered by Nigeria's turn in 1999 toward democratic institutions, disappointment among activists and some members of the U.S. Congress could well lead to a revival of calls for a ban on oil imports and a ban on new U.S. investments in the Nigerian oil sector.

---

129. Executive Order 13129, July 4, 1999.

130. "Pakistan, Turkmens, Afghans to Meet for Pipeline," Reuters, April 20, 1999; and "U.S. Bans Trade with Taliban," *Financial Times*, July 7, 1999.

131. Direct airline traffic between Nigeria's Murtala Mohammed International airport and airports in the United States was prohibited until December 1999, when the U.S. Federal Aviation Agency (FAA) certified that security at the airport met internationally acceptable (International Civil Aviation Association [ICAO]) standards.

132. In April 1994, Nigeria was denied certification as a country that was cooperating sufficiently with the United States in controlling narcotics trade. As a result, the United States is required (by U.S. legislation) to vote against loans to Nigeria in the six multilateral development banks—IBRD, IDA, IADB, ADB, EBRD, and AFDB—listed in the Foreign Assistance Act. Nigeria is also denied OPIC and EXIM assistance and USAID assistance with the exception of certain categories of disaster relief, provision of food and medicines to refugees, and counternarcotics assistance.